IEE CONTROL ENGINEERING SERIES 50

Series Editors: Professor D. P. Atherton
Professor G.W. Irwin

Programming industrial control systems using IEC 1131-3 Revised edition

Other volumes in print in this series:

Volume 2 **Elevator traffic analysis, design and control** G. C. Barney and S. M. dos Santos
Volume 8 **A history of control engineering, 1800-1930** S. Bennett
Volume 14 **Optimal relay and saturating control system synthesis** E. P. Ryan
Volume 15 **Self-tuning and adaptive control: theory and application** C. J. Harris and S. A. Billings (Editors)
Volume 16 **Systems modelling and optimisation** P. Nash
Volume 18 **Applied control theory** J. R. Leigh
Volume 19 **Stepping motors: a guide to modern theory and practice** P. P. Acarnley
Volume 20 **Design of modern control systems** D. J. Bell, P. A. Cook and N. Munro (Editors)
Volume 21 **Computer control of industrial processes** S. Bennett and D. A. Linkens (Editors)
Volume 23 **Robotic technology** A. Pugh (Editor)
Volume 26 **Measurement and instrumentation for control** M. G. Mylroi and G. Calvert (Editors)
Volume 27 **Process dynamics estimation and control** A. Johnson
Volume 28 **Robots and automated manufacture** J. Billingsley (Editor)
Volume 30 **Electromagnetic suspension—dynamics and control** P. K. Sinha
Volume 31 **Modelling and control of fermentation processes** J. R. Leigh (Editor)
Volume 32 **Multivariable control for industrial applications** J. O'Reilly (Editor)
Volume 34 **Singular perturbation methodology in control systems** D. S. Naidu
Volume 35 **Implementation of self-tuning controllers** K. Warwick (Editor)
Volume 36 **Robot control** K. Warwick and A. Pugh (Editors)
Volume 37 **Industrial digital control systems (revised edition)** K. Warwick and D. Rees (Editors)
Volume 38 **Parallel processing in control** P. J. Fleming (Editor)
Volume 39 **Continuous time controller design** R. Balasubramanian
Volume 40 **Deterministic control of uncertain systems** A. S. I. Zinober (Editor)
Volume 41 **Computer control of real-time processes** S. Bennett and G. S. Virk (Editors)
Volume 42 **Digital signal processing: principles, devices and applications** N. B. Jones and J. D. McK. Watson (Editors)
Volume 43 **Trends in information technology** D. A. Linkens and R. I. Nicolson (Editors)
Volume 44 **Knowledge-based systems for industrial control** J. McGhee, M. J. Grimble and A. Mowforth (Editors)
Volume 45 **Control theory—a guided tour** J. R. Leigh
Volume 46 **Neural networks for control and systems** K. Warwick, G. W. Irwin and K. J. Hunt (Editors)
Volume 47 **A history of control engineering, 1930-1956** S. Bennett
Volume 48 **MATLAB toolboxes and applications for control** A. J. Chipperfield and P. J. Fleming (Editors)
Volume 49 **Polynomial methods in optimal control and filtering** K. J. Hunt (Editor)
Volume 50 **Programming industrial control systems using IEC 1131-3** R. W. Lewis
Volume 51 **Advanced robotics and intelligent machines** J. O. Gray and D. G. Caldwell (Editors)
Volume 52 **Adaptive prediction and predictive control** P. P. Kanjilal
Volume 53 **Neural network applications in control** G. W. Irwin, K. Warwick and K. J. Hunt (Editors)
Volume 54 **Control engineering solutions: a practical approach** P. Albertos, R. Strietzel and N. Mort (Editors)
Volume 55 **Genetic algorithms in engineering systems** A. M. S. Zalzala and P. J. Fleming (Editors)
Volume 56 **Symbolic methods in control system analysis and design** N. Munro (Editor)

Programming industrial control systems using IEC 1131-3 Revised edition

R. W. Lewis

r.w.lewis@iee.org.uk

The Institution of Electrical Engineers

Published by: The Institution of Electrical Engineers, London,
United Kingdom

The Institution of Electrical Engineers,
Michael Faraday House,
Six Hills Way, Stevenage,
Herts. SG1 2AY, United Kingdom

British Library Cataloguing in Publication Data

A CIP catalogue record for this book
is available from the British Library

ISBN 0 85296 950 3

Printed in England by Short Run Press Ltd., Exeter

Contents

Preface

The PLC programming languages standard has come from a desire to improve the programming techniques for industrial control systems - not only to enhance software quality but to boost development productivity generally. Although the focus of this book is on the use of the PLC programming languages standard, IEC 1131-3, it is also about the need to change our whole approach to designing control systems. It is time to update our methods and techniques for control system design and start to use modern software engineering practices.

Compared with the computing world, where new programming languages, improved graphical user interfaces and advanced packages for personal computers are being introduced almost every month, change in the world of industrial control systems moves at a much slower pace.

This is quite understandable - with control systems that can manage anything from complex petro-chemical plants, smelting furnaces and automobile production lines to theme-park roller coasters, there is no scope for error - such systems must be robust and reliable twenty-four hours a day. Heavy financial losses and even lives can be at stake if these systems fail to operate as expected.

Innovations and techniques have to prove that they can survive the rigours of industry before they can be accepted. For these reasons, any new fad or technique is viewed with healthy cynicism before being used in anger. So why should anyone take the IEC 1131-3 PLC languages standard seriously? The answer is simple. IEC 1131-3 is not another invention from a standards committee.

The IEC 1131-3 standard is built on well proven techniques which are in use today in different forms in many proprietary control products. For example, the standard includes Ladder Diagram, a standardised form of ladder programming which will be familiar to PLC programmers, and Sequential Function Chart, a graphical sequencing language which is based on Grafcet - a technique that has been in use for a number of years.

The standard has brought many common practices and techniques together to produce a well defined suite of languages. Beyond that, the standard provides a framework for developing well structured control software. With facilities to package proven solutions as function blocks which can be re-used over and over again, the standard can also directly improve productivity - right from the point of analysing an application through to commissioning and long term maintenance.

The first revision of the IEC 1131-3 standard was published in 1993 and is now being implemented by a number of the major PLC manufacturers. It is now forming the basis of the next generation of programmable controllers. I trust that this book conveys part of this new excitement.

Acknowledgments

Firstly, I would like to acknowledge the contributions of all the members of the IEC task force (IEC 65B/WG7/TF3) responsible for drafting the PLC languages standard. Without their diligence, perseverance and sheer enthusiasm, part 3 of the IEC 1131 standard would not exist. Particular thanks go to the chairman, Dr James Christensen of Allen-Bradley, for ensuring that the standard is well written, concise and unambiguous.

Special thanks to David Beaumont of CEGELEC Projects for inviting me to take over his role in the IEC task force as the UK representative and for giving me the opportunity to become directly involved in the development of the standard.

I would like to thank Dr Laurie Burrow of CEGELEC and Paul Roper of Engineering and Technical Services Ltd. for 'volunteering' to check some of the text. I also wish to thank Kloepper and Wiege Software GmbH for the opportunity to review the Multiprog IEC 1131-3 programming system.

I must also mention Dr Mathew Bransby of National Power for his enthusiasm for the standard and permission to use some power station plant specifications as background material for some of the programming examples.

I wish to thank Silvertech Ltd. for material on the use of function blocks in safety related applications.

Finally, I must give special thanks to my family Sue, Martin and Daniel, for leaving me in peace in the attic to get on with the typing!

RWL

Abbreviations and conventions

EEROM	Electrically erasable read only memory
FB	Function block
FBD	Function Block Diagram graphical language
FIP	Fieldbus implementation based on a French standard
HVAC	Heating, venting and air conditioning system
IEC	International Electro-technical Commission
IL	Instruction List language
ISA	Instrument Society of America
ISO	International Standards Organisation
ISP	Integrated Software Project - Fieldbus implementation using existing IEC standards
LD	Ladder Diagram graphical language
MAP	Manufacturing Automation Protocol
MMS	Manufacturing Message Specification
OSI	Open Systems Interconnection
PC	Personal computer[1]
PID	Three term controller, proportional, integral, derivative closed-loop control algorithm
PLC	Programmable logic controller
POU	Program organization unit
RAM	Random access memory
SCADA	Supervisory, control and data acquisition system
SFC	Sequential Function Chart graphical language
ST	Structured Text language

[1] *In the IEC 1131-3 standard, the abbreviation PC refers to a programmable controller. To avoid confusion with the widespread use of PC as personal computer, a programmable controller is referred to as a programmable logic controller or PLC within this book.*

The following font and style is used to distinguish examples of textual programming language from ordinary text :

```
BEGIN
        A := A + 100.5;  (* Comment *)
END
```

The suppression of superfluous detail in textual language examples is indicated using an 'ellipsis' as follows:

```
...
```

Note: From 1 January 1997, the IEC has changed the numbering scheme for industrial standards. Some standards which formerly had a four digit identifier have been renumbered by prefixing the identifier with a '6'. The full identifier for the PLC standard, formerly IEC 1131 is now IEC 61131. Similarly, the IEC Function Block standard IEC 1499 has become IEC 61499. For brevity, the original four digit numbering scheme has been retained in this book.

Chapter 1

Introduction

Using this book

This book has been written as a user guide and tutorial for the new international standard IEC 1131-3 for programmable logic controllers (PLCs). If you are not familiar with the background and purpose of the standard it is recommended that you read through this introductory chapter.

If you are considering designing or writing control programs using any of the IEC 1131-3 languages for the first time, you are advised to gain an understanding of how programs are structured and can be built up from various software elements; this is reviewed in Chapter 2, 'IEC 1131-3 concepts'. This chapter will also explain how the standard provides a selection of programming languages for developing different parts of a control program.

Later chapters give specific details on using each of the IEC 1131-3 languages. The suitability of each language for solving different types of industrial control problems is described along with techniques and examples.

Facilities in the IEC 1131-3 and an accompanying standard IEC 1131-5 for communicating with other PLCs and devices are discussed in Chapter 12, 'Communications'.

Although it is the intention that this book should accurately describe the languages and concepts of the standard, you should always refer to the published IEC 1131-3 standard for all formal definitions of any part of the standard. Alternative interpretations are described in the text in situations where some minor details in the standard are ambiguous.

1.1 The start of a new generation

Every industry reaches a degree of maturity when there is a need to consolidate different designs and approaches. Inevitably a point is reached where standards start to emerge by various means. Standards may evolve through de-facto use of a particular technique or approach. In other situations, a standard may be foisted on the industry through statutory regulations.

Consider any industry that involves building systems or providing services using different components, where the components are manufactured in different locations, different companies or even different countries. There is clearly always a need to have standards of some form for the disparate components to work together harmoniously. Progress with many of the early industries of the industrial revolution was stalled until standards were developed. Railway networks could only grow when a standard gauge was adopted; public power utilities could only start to distribute electrical power efficiently with the acceptance of standard supply voltages and AC frequencies.

The programmable logic controller (PLC) market has reached such a turning point. During the last ten to fifteen years a wide range of different programming techniques has been used to write programs for industrial control applications. Control applications have been developed in BASIC, FORTH, C, Structured English, Instruction List and numerous other proprietary languages.

One particular programming technique, ladder logic, has gained wide acceptance because its understanding is fairly intuitive. Although it has been adopted by most PLC manufacturers, each company's implementation uses its own particular dialect. As a result, unfortunately, the only thing that can be said of all of these programming languages is that they are all different.

For people involved with such systems, from technicians, maintenance personnel and system designers to plant managers, this results in inefficient use of time and money. If plant personnel or a company has to work with a number of different control systems, there is clearly a waste of human resources involved in training staff in skills in many different control languages.

Having so many PLC languages can also result in misunderstandings which, in some cases, could have disastrous results. Just consider the maintenance engineer who, in the middle of the night, is requested by a harassed plant manager to diagnose rapidly a safety interlock fault on a production line because the line has stopped. The system may involve two PLCs made by different vendors. In this situation, subtle differences in interpreting the two ladder diagram programs could lead the engineer to a faulty diagnosis and possibly to making a hazardous modification to one of the programs.

Fortunately, the international industrial community has recognised that a new standard for programmable logic controllers is required. A working group within the International Electro-technical Commission (IEC) was set up in 1979 to look at the complete design of programmable logic controllers, including the hardware design, installation, testing, documentation, programming and communications. The IEC is a sister organisation to the International Standardisation Organisation (ISO) based in Geneva and has committees and working groups formed from representatives put forward by standardisation bodies of most industrial countries of the world.

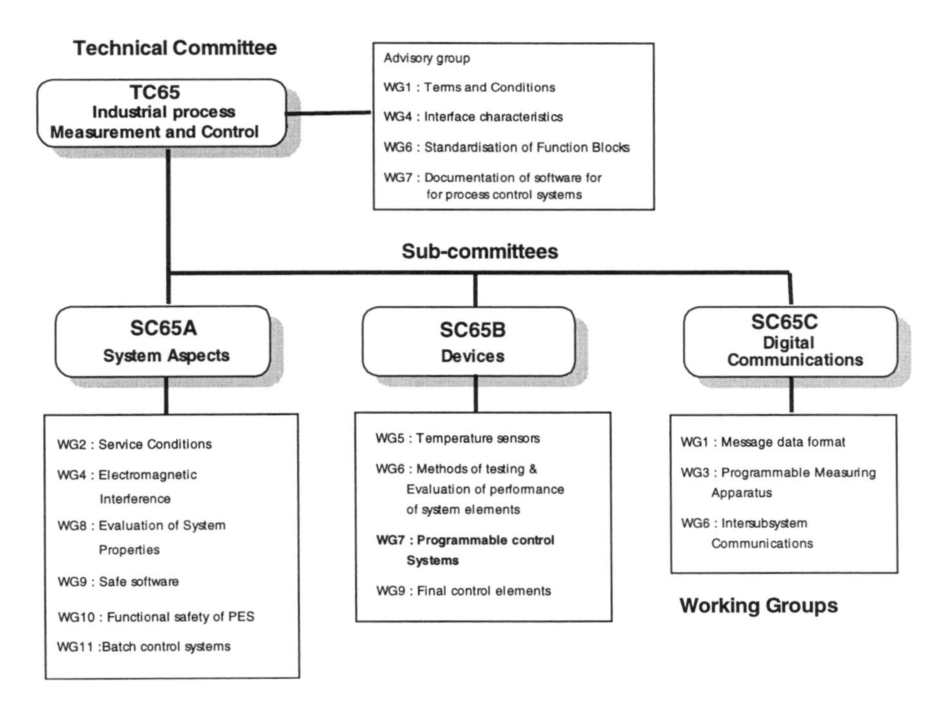

Figure 1.1 IEC Working Group structure 1993

Note: This book was revised in Autumn 1998 to reflect the corrections and amendments proposed by the IEC for the second edition of IEC 1131-3. The second edition is planned to be published in late Summer 1999.

IEC 1131-3 Technical Report Type 2 — This report contains proposals for possible future enhancements and extensions for a third edition of IEC 1131-3 to be developed after 2000.

The IEC technical committee TC65, which is concerned with standards related to industrial-process measurement and control, established Working Group 7 to develop the PLC standard. This working group, which is now designated IEC 65B/WG7, quickly realised that standardisation of all of the issues regarding PLC design was beyond the capability of a single group. A number of specialist task forces were therefore established to develop different parts of the standard. Task Force 3 has had the primary objective of developing a new language standard, which has become part 3 of the IEC PLC standard 1131.

Throughout this book part 3, the PLC Languages standard, will be referred to simply as IEC 1131-3. Figure 1.1 depicts the structure of the IEC standardisation groups in existence in 1993 when the 1131-3 standard was first published.

Although this book is primarily concerned with PLC software, Working Group 7 is developing a complete suite of standards that addresses all aspects of PLCs and their application to industrial-process measurement and control. The various parts of the full IEC 1131 standard are listed in Table 1.1.

Table 1.1 Parts of the IEC 1131 standard

Part	Title	Contents	Published
Part 1	General information	Definition of basic terminology and concepts.	1992
Part 2	Equipment requirements and tests	Electronic and mechanical construction and verification tests.	1992
Part 3	Programmable languages	PLC software structure, languages and program execution.	1993
Part 4	User guidelines	Guidance on selection, installation, maintenance of PLCs.	1995

Table 1.1 Continued

Part	Title	Contents	Published
Part 5	Messaging service specification	Software facilities to communicate with other devices using communications based on MAP Manufacturing Messaging Services.	Planned publication 1998
Part 6	Communications via fieldbus	Software facilities of PLC communications using IEC fieldbus.	Awaiting completion of fieldbus standards
Part 7	Fuzzy control programming	Software facilities, including standard function blocks for handling fuzzy logic within PLCs.	1997
Part 8	Guidelines for the implementation of languages for programmable controllers	Application and implementation guidelines for the IEC 1131-3 languages.	Planned publication 1998

1.2 The growing PLC market

The application of this standard is set to have a major impact on the rapidly growing industrial process control and instrumentation market, a market which has grown significantly over the last ten years. It is estimated that this market in Europe alone, while still in recession, was worth $1.52bn in 1995 and will continue to rise by 5% per annum to reach $2.14bn by 2002. A global compound market growth rate of at least 5% is possible well past 2002. As PLCs are the backbone of most automation projects of the process control and instrumentation sector, the PLC market can be expected to increase by an even greater proportion. With many of the functions that were previously offered in separate instruments, such as closed loop PID control, now being integrated into PLCs, the PLC market can be expected to expand in many new directions.

The PLC market can be differentiated by size into: micro PLCs with up to 100 Input and Output (I/O) points, small PLCs with between 100 and 200 I/O points, medium PLCs with 250 to 1000 I/O points and large PLCs with upwards of 1000 I/O points. It is in the micro and small PLC market sectors that the biggest growth is expected.

Most industrial sectors use PLCs; however the main applications that are expected to increase include food and beverage, power generation, water, glass, cement, plastics and rubber. The largest share will however continue to be in chemical and petrochemical sectors. In Europe, the introduction of stringent safety or environmental regulations is causing industries such as utilities and water treatment to upgrade their control systems.

The re-equipping of much of the industrial base of Eastern Europe, particularly in heavy industry, steel production and processing of staple food products, will also require an inevitable increase in the demand for PLC based control systems.

In the long term, the de-skilling and automation of production lines and manufacturing processes will lead to the development of the 'lights-out' factory where there is no longer any significant requirement for production workers or operators. The complete 'lights-out' factory will behave as a single machine dedicated to the production of a range of products. Such levels of automation will not be achievable without a large investment in control systems, particularly PLCs.

In a large number of these industrial sectors and future applications, there will be an almost continuous demand for cost savings through improved manufacturing efficiency and process optimisation.

This means that PLCs will be required to integrate readily with other systems and devices and be sufficiently flexible that the control strategies for new processes can be quickly brought on stream. The maturing industrial control market is now starting to ask for 'open system' products which can be easily integrated into both manufacturing cells and plant-wide computer systems.

1.3 Towards 'open systems'

The term 'open system' is probably one of the most abused of all those found in computer marketing literature and unfortunately has become somewhat devalued. However, in principle, an 'open system' product is one that adopts current industrial standards and techniques so that integration with other 'open system' products is straightforward and without undue trauma. There are few standards suitable for industrial systems that actually address all the requirements for developing complex industrial applications.

The most notable attempt to bring harmony into an industry so full of disparate control devices, instruments, communications protocols and PLCs was started by General Motors (GM) in the USA with the development of the MAP (Manufacturing Automation Protocol). In the mid-1980s, there were few common methods available for connecting control devices from different vendors so that they could communicate sensibly.

The integration of different devices usually meant the development of complex software in the form of communications drivers. A driver translates between communications messages in one format sent by one device and messages of a different format as expected by the other device and vice versa, a technique known as protocol conversion. GM realised that a large proportion of the time and cost of building large systems, particularly for automating car production lines, could be saved by having a common communications standard.

The MAP standard has now stabilised but has not been particularly successful for a number of reasons. The hardware costs to provide a communications interface from a PLC into a MAP network can be very high and the end-to-end communications performance has not been very impressive. Even when a fully compliant control device such as a PLC is connected to a MAP network, a significant problem still remains: how to transfer meaningful information between devices which have probably been programmed in different languages.

The Instrument Society of America (ISA) standardisation work to develop a Fieldbus standard for the interconnection of distributed field devices such as pressure sensors, temperature controllers and valve positioners has taken a long time to come to fruition. Although Fieldbus does address some of the issues regarding the structure of software within devices, the Fieldbus standard is not proposing any particular techniques or languages for developing programs within devices:

To summarise, the industrial instrumentation and control systems need an 'open systems' approach to build large systems using equipment from different manufacturers. Until the IEC 1131-3 standard was published in March 1993, there was no suitable standard that defined the way control systems such as PLCs could be programmed. Communications standards only addressed how information could be exchanged.

1.4 Deficiencies of current PLC software

Apart from the fact that PLCs are all programmed differently, there are a number of notable problems when trying to program complex control applications using many of the current PLC languages.

1.5 Ladder programming

Ladder programming has evolved from the electrical wiring diagrams that are used in the car industry for describing relay control schemes. For small systems ladder programming has served the industry well. The basic programming skills needed to develop small applications using ladder programs can be learnt relatively quickly and the graphical presentation can be understood almost intuitively. The technique is particularly easy to understand by people who are familiar with simple electronic or electrical circuits. Consequently, it is well accepted by electricians and plant technicians.

For the maintenance engineer, this programming technique is also very popular because faults can be quickly traced. Most PLC programming stations generally provide an animated display that can clearly identify the live state of contacts in rungs of a ladder program while a PLC is running. This provides a very powerful on-line diagnostic facility that can clearly show up any logic paths through ladder rungs that are incorrect.

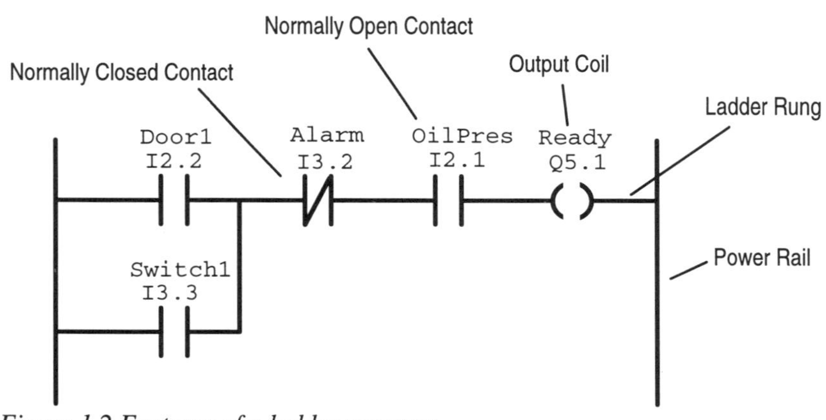

Figure 1.2 Features of a ladder program

With the increase in memory size of many of today's PLCs, ladder programming can now be used to build very large systems, bringing significant software development problems. It is often very difficult to build well structured programs using ladder programs because generally there is limited or no support for procedures, subroutines or program blocks.

In this simple example, an output coil 'Ready' is powered, i.e. is set on, when contacts 'Door1' or 'Switch1' are made, contact 'Alarm' is off, and contact 'OilPres' is on.

Weak software structure

Most PLCs offering ladder programming support a limited number of subroutines or program blocks. It can be difficult to break a complex program down hierarchically. There are generally very limited facilities for passing parameters between program blocks. As a result it is hard to break a large program down into smaller parts that have clear interfaces with each other. Consequently, as it is usually possible for one part of a Ladder Diagram program to read and set contacts and outputs used in any other part of the program, it is very difficult to have truly encapsulated data.

Encapsulation, the ability to hide internal information within program blocks so that it cannot be interfered with by other parts of a program, is very important for creating good quality, maintainable software.

The lack of data encapsulation can be a serious problem with large programs especially if they are written by different programmers. Each programmer must always be sure that only certain global registers and memory areas are modified when communicating information between different parts of a large program. There is always a danger that an internal data register of one program block is modified by faulty code in another block. Clearly, with PLCs controlling large plant and machinery, such faults could result in unpredictable and catastrophic control behaviour.

Low levels of software re-use

Some large programs may use the same logic strategy or algorithm over and over again. For example, consider a program for PLC based fire and gas safety systems for a large petro-chemical plant. Such systems may use the same logic strategy for detecting the presence of fire in hundreds of different plant areas using a fire detection voting algorithm. Because fire detectors are notoriously unreliable, there is usually a requirement for at least two detectors in an area to register a fire before extinguishants are released.

Such systems are usually written by repeating the same set of ladder rungs hundreds of times with just minor modifications to read different input values or set different outputs. This can result in very large and unmanageable programs.

Writing repeated logic in a block that can be called many times can greatly reduce the size of programs and produce programs that are easier to maintain. In many ladder program based PLCs, facilities for re-using software in this way are usually limited and difficult to use.

Poor data structure

There are also problems using structured data with ladder programs. Typically data is stored and addressed in single memory bits or registers. Using a single bit is fine for holding the status of simple flags and digital inputs and outputs. Similarly, registers, which generally hold 16 bits, can be used for counters and analogue values. However, with many complex applications, there is usually a requirement to group data together as a structure. PLC ladder programs do not normally have any facilities for structured data and yet it is a standard feature of most high level languages such as PASCAL and C.

For example, consider a pressure sensor that is connected to a PLC digital input that detects when a certain threshold pressure is reached. The PLC hardware I/O management system will create a single bit within the PLC memory that reflects the state of the connected sensor. However, a single bit may be insufficient to reflect all of the states of the sensor. The application may require that the sensor can be disabled and placed in a test mode, that a time is recorded when the sensor becomes active, that a sensor alarm is raised when the sensor is 'on' for more than a prescribed period and so on.

All of this extra information associated with the sensor should ideally be stored as a single structure that can be addressed using a common unique name. With complex ladder programs, such data are often spread out throughout the PLC memory so that the chances of accidentally addressing data bits associated with a different sensor are very high. With no formal data structure support, the PLC programming system is generally unable to give the programmer any warning when such data violations occur.

Limited support for sequencing

Many industrial applications require that the PLC programs generally require one or more sequences of operations depending on various operational states of the controlled process.

For example, consider starting a conveyor belt for a production line: the program may need to check that certain auxiliary equipment is in a known initial state; it may then need to assign speeds to the conveyor drive motors as a series of steps. After each speed change, the PLC may need to check that the speed of the whole line has stabilised before making the next speed step. When the working speed is reached, it may need to activate other machines and equipment on the line.

The conventional approach to this type of problem using ladder programming is to assign a bit to each state of the sequence and to each transition that triggers a change from one state to the next. The sequence is programmed in three parts: (1) ladder rungs for detecting the transitions from one state to the next, (2) rungs switching between states and (3) rungs to perform actions or operations when particular states are active. Figure 1.3 depicts part of a ladder program for solving the conveyor start-up example.

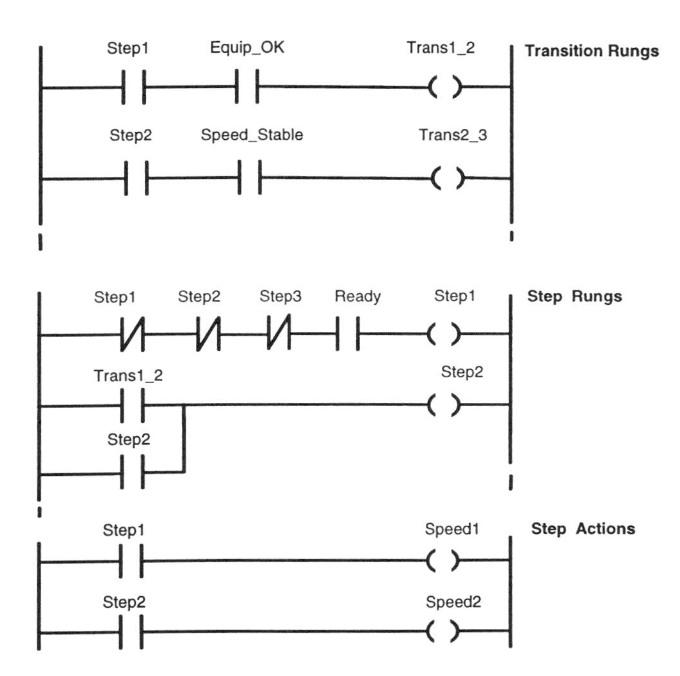

Figure 1.3 Sequencing using ladder programming

It is clear that ladder programs can produce sequential behaviour in this way; however with large programs which may have sequences with hundreds of steps, this approach can quickly become very unwieldy and difficult to maintain. The main problem is that the control of the sequence is mixed in with the application logic so that the overall behaviour is difficult to understand.

Note: Many PLC vendors are now offering Grafcet or Sequential Function Charts from the IEC 848 standard; we will see that this is a superior method of defining sequences. Sequential Function Charts are part of the IEC 1131-3 standard and will be discussed in some detail in later chapters.

Limited execution control

Figure 1.4 depicts the way ladder programs are normally executed. Most PLCs continuously execute a cyclic scan. The cycle starts with the hardware I/O system gathering the latest values of all input signals and storing their values in fixed regions in memory.

The rungs of the ladder program are then executed in a fixed order starting at the first rung. During the program scan, new values of physical outputs, as determined from the logic of the various ladder rungs, are initially written to an output memory region. Finally, when the ladder program has completed, all the output values held in memory are written out to the physical outputs by the PLC hardware in one operation.

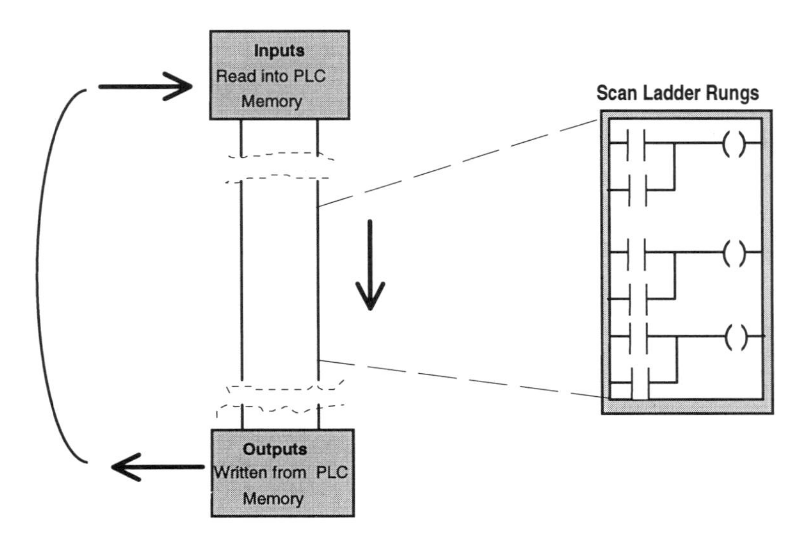

Figure 1.4 Scanning ladder programs

This simple execution scheme is easy to understand and is quite adequate for a wide range of elementary PLC applications. However there are a number of problems with this approach. Generally, the frequency of the ladder program execution depends to a large extent on the length and complexity of the ladder program. Add more rungs and the scan time will lengthen; this can cause problems when designing systems that need to respond within a guaranteed period.

For example, consider a requirement of a PLC system to control a plastics extrusion machine. The PLC may need to control motor drives for handling the plastics feed screw and extrudate haul-off, while at the same time controlling mechanical interlocks and operator displays and managing the delivery of plastics feedstock.

To protect the expensive extruder barrel, the PLC should rapidly activate motor drive trips within, say, 100 milliseconds if certain alarm signals such as a 'melt over-pressure' are detected. The pressure of molten plastics or the melt pressure is a critical parameter in an extruder and may suddenly increase due to some form of blockage further along the barrel. A high internal pressure may cause the barrel to distort or even explode.

In this case, critical sections of the ladder program detecting the alarm conditions will need to execute within, say, 50 milliseconds to always meet the 100 millisecond deadline. It may be acceptable for long sections of the ladder program handling the other less critical extruder functions to run less frequently. Consequently, the programmer may need to spend time and effort arranging the scheduling of the different parts of the program to meet these deadlines.

Generally this can be achieved by running the faster logic sections every 50 milliseconds and switching in different sections of the less critical ladder rungs in the background.

Although most PLCs provide some facilities to select when various sections of rungs are scanned, such as the master control relays in Allen-Bradley PLC-5s, it is generally very difficult for the programmer to have full control over the execution rates of different parts of the program.

When PLC programs are also used for running closed loop control algorithms such as the three term PID (proportional, integral, derivative), the problems are compounded. For good control, the period between updates of the PID algorithm is critical, i.e. the time between the ladder program scans should be stable and generally of a known duration.

This may be difficult to realise in practice. There is always a possibility that just adding a few new rungs or modifying a ladder program can change the scan execution time. This may subtly change the timing of certain closed loop control algorithms so that the whole system behaves differently. This lack of *deterministic* behaviour is one of the main criticisms of using conventional ladder programs in PLC applications for process control.

We will see that the IEC 1131-3 standard has given some attention to this problem and has provided comprehensive execution control of different program elements.

Cumbersome arithmetic operations

Finally, performing arithmetic operations using ladder programs can also be very cumbersome. There is a wide range of different techniques available but like ladder programming itself, the approach tends to vary from PLC product to product. One common approach is the use of arithmetic blocks. Such blocks can be triggered by the activation of a ladder rung. Figure 1.5 depicts arithmetic blocks that might be used to calculate the average of two analogue values held in registers.

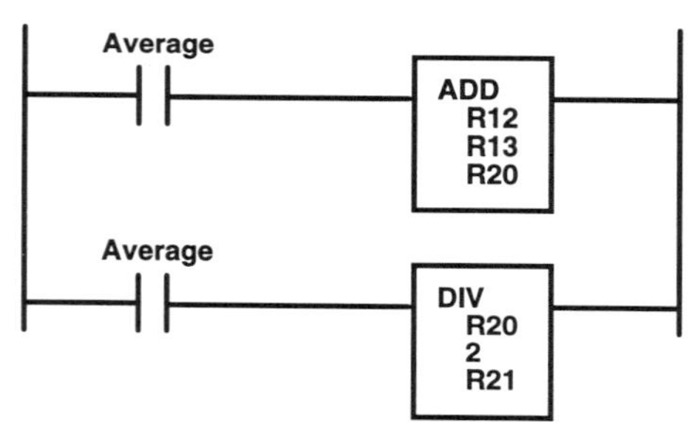

Figure 1.5 Arithmetic operations in ladder programs

As with the other ladder programming features, performing simple arithmetic operations in this way is straightforward. Trying to perform complex calculations involving dozens of variables can be a nightmare and can result in programs that are very difficult to maintain.

Because PLC programming stations generally only allow a few ladder rungs to be visible at a time, it is only possible to see a small part of a complex calculation at a time. As a PLC programmer remarked, ‘It's like trying to write a program through a small key-hole’.

1.6 The deficiencies of conventional ladder programming

The following table summarises the main faults with conventional ladder program based PLCs.

1. **The ladder symbols and facilities vary between different PLC products.**
2. **Poor facilities for structured or hierarchical program decomposition.**
3. **Limited facilities for software re-use.**
4. **Poor facilities for addressing and manipulating data structures.**
5. **Limited facilities for building complex sequences.**
6. **Limited control over program execution.**
7. **Facilities for arithmetic operations are cumbersome.**

For people who work with PLCs every day, this summary may seem to be somewhat harsh. With very good programming standards, use of program generators, off-line documentation tools and the application of information technology (IT) using databases to manage data addressing and so on, many of these deficiencies can be circumvented to some degree but never completely ignored.

Considering the wide range of PLC applications from the control of automobile production lines, brewing, water treatment and fun-fair roller-coasters to treatment of radioactive waste, it is remarkable that quality systems can be delivered despite these limitations. But at what cost and effort ?

1.7 IEC 1131-3: improving software quality

One of the major objectives of the new IEC 1131-3 PLC programming standard is to widen the understanding of PLC programs. With PLCs being used in so many industries, and with many applications concerned with safety, it is important that programs, and subsequently the behaviour of controlled plant, can be understood by a wide range of industrial personnel. A PLC program should be relatively easy to follow by technicians, plant managers and process engineers alike, as well as by the programmer.

By having a consistent approach to programming, the standard will also of course reduce the learning-curve for all people who need to deal with PLCs - especially if they have to deal with products from different vendors.

Before detailing the new features in the standard it is worth considering all the attributes of software and the software development process that contribute to improvements in software quality. These quality attributes are described in part 8 of the IEC 1131 standard that provides guidelines on the application and implementation of IEC 1131-3. These attributes can be used as part of a quality checklist for any piece of software and are particularly relevant to PLC software. We will see that many of the features of the IEC standard directly improve many of these quality attributes of PLC software.

Capability

This describes the extent to which a system can perform its intended design function. This is influenced by factors:

Responsiveness	The time required for the system to produce appropriate responses to a specified combination of external events.
Processing capacity	The extent to which the system can meet scheduling deadlines under specified sets of conditions.
Storage capacity	The extent to which the system can retain in memory all the required programs and data under specified sets of conditions.

Availability

This describes the proportion of time in the life of a system when it is available for its intended function. This is influenced by factors such as:

Reliability	The ability of the system to continue to perform all its intended functions over a specified period of time and range of conditions. An inverse measure of reliability is mean time between failures (MTBF).
Maintainability	The ease with which the system can be restored to full capability after the occurrence of one or more faults from a specified set. An inverse measure of maintainability is mean time to repair (MTTR); availability is often defined by the expression MTBF/(MTBF+MTTR).
Integrity	The degree to which the system can continue to perform all its intended functions over a specified range of threats, including both unintended user actions and intentionally hostile actions.

Usability

This describes the ease with which a specified set of users can acquire and exercise the ability to interact with the system in order to perform its intended functions. This is influenced by factors such as:

Entry requirements	The amount of formal and informal training required before the user can learn to interact with the system (e.g. prerequisite educational level, skill-sets in the use of operating systems, windows, menus, etc.).
Learning requirements	The training required for a user meeting a specified set of entry requirements, to learn to interact with the system to perform a specified set of system functions.
User productivity	The number of system-related operations per unit time which can be performed by a user with a specified level of training and experience.
Congeniality	The extent to which a user prefers to utilise the system software to perform the intended system functions, with respect to alternative activities that may or may not be system-related. In other words - is the system *user friendly* ?

Adaptability

This describes the ease with which the system may be changed in various ways from its initial intended functions. This is influenced by factors such as:

Improvability	The ease with which the existing *capability*, *availability*, and/or *usability* can be upgraded without significantly changing the system functionality.
Extensibility	The ease with which new functionality can be added to the system.
Portability	The ease with which system functionality can be moved from one system to another.
Reusability	The ease with which the functional capabilities of an existing software element can be used in a new or in a different system.

It is the intention that IEC 1131-3 based PLC software will have significantly better software quality attributes, as described, than software generated for conventional PLCs. This is achieved by new language features and providing different programming languages for solving different types of industrial control problem.

1.8 The main features of IEC 1131-3

The IEC working group responsible for the 1131-3 standard started by reviewing commonly used techniques and languages as offered by various PLC manufacturers. They then rationalised the different languages and numerous dialects of ladder programming to create a new set of languages. Considering that the major PLC companies including Allen-Bradley and Siemens have active members in the IEC group, it is a notable achievement that the group has been able to come to a consensus and produce a robust and working standard.

The major new features offered by IEC 1131-3 are summarised as follows:

1. The standard encourages **well structured**, 'top-down' or 'bottom-up' program development. The standard allows a program to be broken down into functional elements that are referred to as program organisation units or POUs. POUs include functions, function blocks and programs. These will be described in detail in later chapters. Any IEC software element can be constructed hierarchically from other more primitive elements.

2. The standard requires that there is **strong data typing**. This means that the PLC program station should be able to detect when a programmer erroneously attempts to write the wrong type of data to a variable. For example, assigning the value 1 as a bit, to a memory location designated to receive the value of a time duration as 1 second, would be detected as an error before the program is down-loaded into the PLC. This will remove a major source of programming errors in conventional PLC ladder programs.

3. Facilities are provided so that different parts of a program can be executed at different times, at different rates and in parallel, i.e. there is support for **full execution control**. For example, part of a PLC program may need to scan every 50 ms to ensure that all changes of state read in from mechanical limit switches are detected, while another part of a program monitoring temperature sensors may only need to scan every second. IEC 1131-3 provides facilities so that the programmer can decide to run these different parts at different rates, by assigning the parts to different tasks.

4. There is full support for describing sequences so that **complex sequential behaviour** can be easily broken down using a concise graphical language called Sequential Function Chart. This allows a sequence to be described in terms of steps, their actions and transitions between steps. There is full support for sequences that call alternative sequences and for sequences to call other sequences that run in parallel. For example, a PLC program may be required to control two independent reactor vessels. After the PLC program has initialised, the sequences handling the two vessels can be programmed to run in parallel and can then run independently of each other.

5. There is support for defining **data structures** so that data elements that are associated can be passed between different parts of a program as if they were a single entity. For example, a pump controlled by a PLC may have a status bit defining whether it is operational or not, an analogue output defining the required speed setting, an analogue input defining its measured speed and so on. All of these data elements can be defined as in a single 'pump' data structure. It is then possible to pass all of this information as a single variable between different program organisation units. This improves program readability and ensures that associated data is always accessed correctly.

6. The standard provides **flexible language selection**, i.e. a set of three graphical and two textual languages for expressing different parts of a control application. The system designer is free to choose the language that is most

suitable to solve a part of an application program. Different parts of a program can be expressed in any of the languages.

7. IEC 1131-3 provides standardised languages and methods of program execution so that a wide range of technological problems can be programmed as **vendor-independent software** elements. In other words, a large proportion of software written for IEC 1131-3 compliant PLCs should be portable and will run on PLCs from different vendors.

The new IEC 1131-3 languages that will be described fully in later chapters are Structured Text, Function Block Diagram, Ladder Diagram, Instruction List and Sequential Function Chart.

Structured Text (ST) - A high level textual language that encourages structured programming. It has a language structure (syntax) that strongly resembles PASCAL.[1]

```
IF SPEED1 > 100.0 THEN
    Flow_Rate := 50.0 + Offset_A1;
ELSE
    Flow_Rate := 100.0;  Steam := ON;
END_IF;
```

Function Block Diagram (FBD) — A graphical language for depicting signal and data flows through function blocks — reusable software elements.

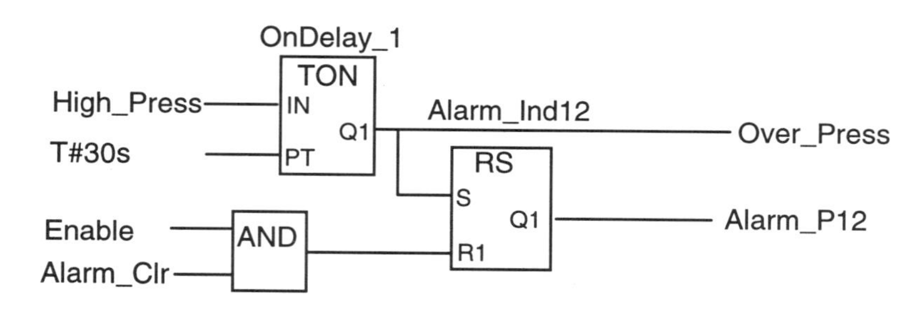

[1] *Some features of the IEC textual languages have also been influenced by ADA, a high level language designed for the American Department of Defence (DOD), 1979.*

Ladder Diagram (LD) — A graphical language that is based on relay ladder logic, a technique commonly used to program current generation PLCs.

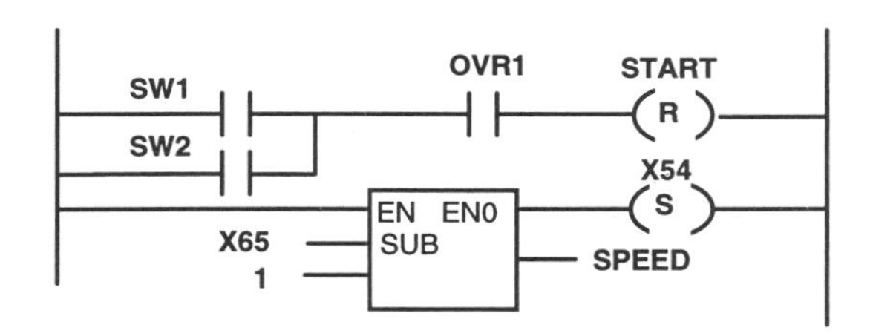

Instruction List (IL) — A low level 'assembler like' language that is based on similar languages found in a wide range of today's PLCs.

```
        LD    R1
        JMPC  RESET
        LD    PRESS_1
        ST    MAX_PRESS
RESET:  LD    0
        ST    A_X43
```

Sequential Function Chart (SFC) — A graphical language for depicting sequential behaviour of a control system. It is used for defining control sequences that are time- and event-driven.

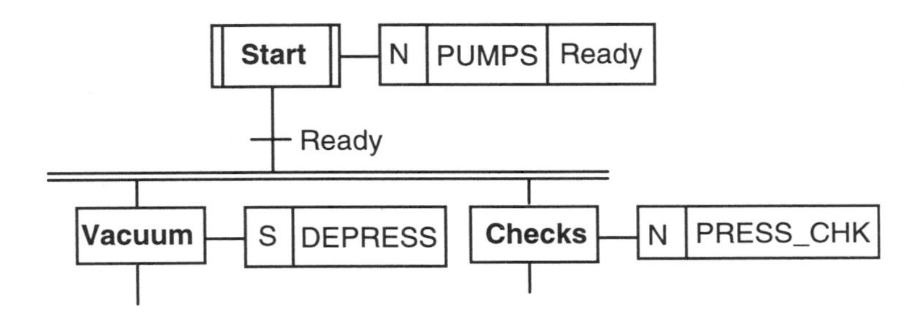

In addition to the new languages, the IEC 1131-3 standard also formally defines a number of powerful new concepts, notably the use of function blocks.

1.9 Function blocks — software integrated circuits

We have already said that IEC 1131-3 encourages the development of well-structured software that can be designed either from top-down or bottom-up. One of the main features in the standard to support this is the use of function blocks.

A function block is part of a control program that is packaged so that it can be re-used in different parts of the same program, or even in a different program or project. A function block may provide the software solution to some small problem, such as creating a pulse timer, or may provide the control to a major piece of plant or machinery, for example the control of a pressure vessel.

The ability to store data as well as the algorithm within the function block gives the IEC function block significant advantages over similar concepts found in other languages, such as subroutines in FORTRAN and functions in C. A function block describes both the behaviour of data and the data structure.

The IEC standard ensures the use of well defined interfaces into function blocks, i.e. that there are formal definitions for each input and output parameter. Function blocks designed by different programmers can therefore be readily interconnected because each input and output parameter must use data types derived from data types defined within the standard.

Function blocks should be regarded as the basic building blocks of a control system. The standard provides facilities so that well defined algorithms or control strategies written in any of the IEC languages can be packaged as re-usable software elements. Function blocks not only encourage well structured design, but in the long term can greatly speed up the application development. This is especially true when developing similar applications where high re-use of standard solutions, as function blocks, should be possible.

Through the increased use of function blocks, we may start to see the development of 'vendor-neutral' industrial software; in other words, markets emerging for libraries of function blocks designed for solving a wide range of different industrial applications that can run in any manufacturer's product.

Although IEC 1131-3 defines a few rudimentary function blocks, there are currently no plans to define a range of standard industrial blocks for, say, signal ramps, filters, PID control, etc. However, in the current work on Fieldbus standardisation, the idea of having a standard range of blocks that can interoperate across a fieldbus network between different vendors' products is being pushed very strongly.

Comparisons have been made between function blocks and objects found in object oriented programming. Function blocks contain encapsulated data, and have methods associated with that data, i.e. the algorithm or control strategy. However, other features found in object oriented languages such as inheritance and polymorphism are not supported.

The function block concept may be more clearly understood by hardware analogy. In many ways, function blocks can be compared with integrated circuits (ICs) or 'silicon chips' which are used every day by electronic engineers. The use of ICs has clearly revolutionised the design of electronic hardware - no one would now consider building a system up from primitive circuits unless there are some very special reasons. Like ICs, function blocks can provide 'off-the-shelf' solutions to common control problems and so could be set to revolutionise the development of control software.

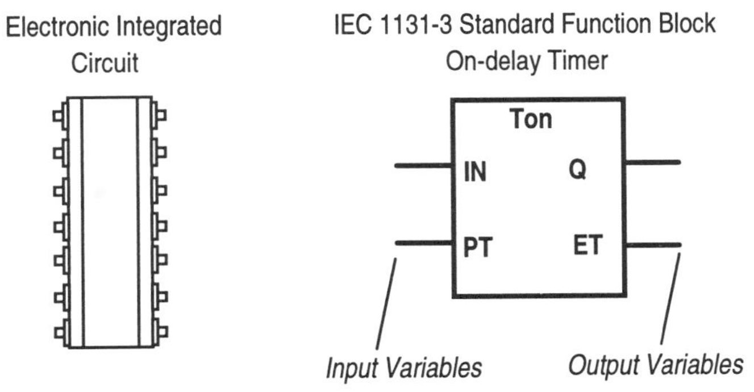

Figure 1.6 Comparison between ICs and function blocks

The idea of considering re-usable software as ICs was first put forward by Brad Cox in 1986 [2]. As the world began to accept the object oriented approach for mainstream software development, he predicted that software vendors would start to market libraries of useful software objects.

Unfortunately, this has not developed in quite the way that he foresaw, mainly because there is such a wide variety of different object oriented languages. Objects developed for one language are not directly compatible with any of the others, e.g. C++, Objective C, Smalltalk 80. However, a market has recently developed for object libraries (as software components) for Microsoft's Visual Basic[3]. If IEC 1131-3 products become the norm, it is possible to foresee markets for application specific function block libraries developing, for example, for PID algorithms, fuzzy logic controllers, alarm handlers and so on.

[2] *"Object oriented programming - An evolutionary approach", Brad J. Cox, 1986*

[3] *Visual Basic is a trade mark of Microsoft Corporation.*

1.10 Software encapsulation

Packaging software using *encapsulation* and *information hiding,* so that software internals are not readily assessable, is important in improving software quality. Encapsulation is the ability to handle a collection of software elements as a single entity, for example packaging a number of data items in a single data structure.

Table 1.2 IEC 1131-3 software elements

IEC element	Description	Encapsulation		Hiding	
		Data	**Procedure**	**Data**	**Procedure**
Structure	Used for forming a new composite from a number of data types	**Yes**	**No**	**No**	**No**
Function	Used for calling re-usable software procedures	**No**	**Yes**	**Yes**	**Yes**
Function block	Used to package software procedures and data	**Yes**	**Yes**	**Yes**	**Yes**
Program	Used to build entire IEC programs from other elements	**Yes**	**Yes**	**Yes**	**Yes**
Action	Used to package a set of operations for use in sequence steps	**No**	**Yes**	**No**	**No**
Access path	Used to allow access to one or more items of data from a remote device	**Yes**	**No**	**Yes**	**No**

'Information hiding' is the restriction of external access to data and procedures within an encapsulated element. For example, in a large control project, consider the benefits of making parts of a well tried and tested control strategy into function blocks, each with a small set of external parameters providing all the required program control. System designers are then able to build up larger applications simply by connecting the proven blocks — intimate knowledge of how the block has been coded internally is not required.

It is only necessary to understand the 'functionality' provided by a function block and use of its external parameters. As the system designer is only able to change the values of external parameters, the danger of using a packaged control strategy with conflicting or inconsistent data and control signals is very much reduced.

In contrast, consider the situation where the re-usable code is not packaged as a function block, and the system designer is able literally to copy the code from previous control solutions. There is now every danger that the code will be modified or that its internal data are accidentally over-written. The chances of the code behaving as originally designed and tested in this situation are significantly reduced.

Apart from function blocks, IEC 1131-3 provides a number of language elements for encapsulating both software procedures and data. These will be discussed in more detail in further chapters of this book. Table 1.2 lists the main software elements in the IEC standard. The table emphasises that data and code (procedure) encapsulation and information hiding are characteristics of many of the IEC software elements.

1.11 Inter-system portability

A major benefit for end-users using IEC 1131-3 compliant products will be the possibility of being able to port control system software to different products. Software developed for one manufacturer's PLC product family should theoretically be able to run on any other IEC compliant system. Clearly, this could have a major impact on sourcing PLC equipment and will generally open up the PLC market.

The choice of a PLC supplier for a new project should be determined more by the suitability of the hardware, I/O sub-system and price than by the types of languages that the PLC supports.

Such a high level of software portability, referred to as *inter-system portability*, may be difficult to realise in practice. The IEC 1131-3 standard defines numerous tables of features and only requires that PLC vendors list those features that they decide to implement. In other words, the IEC 1131-3 standard does not require or expect every language feature to be implemented!

The normal chances then of finding two vendors who will choose to implement exactly the same set of features may not be very high. Without products having the same set of language features, the possibility of being able to port software between such systems will be slight.

As the IEC standard is not explicit about compliance requirements, any PLC vendor can claim to be IEC 1131-3 compliant by simply implementing one or two simple language features. Because of IEC 1131-3's weak compliance requirements and the possibility that the dream of truly portable control software was in jeopardy, a number of companies concerned with promoting IEC 1131-3 based products formed the PLCopen trade association. PLCopen has been actively involved with defining different compliance levels with the objective that products that have the same compliance level will support a known level of software portability.

It is worth noting that membership of PLCopen has continually grown since it was formed in 1992 illustrating the widening international impact of IEC 1131-3. At the end of 1993 it had members from over 41 different companies and organisations. By 1997 it had almost 100 members representing most of the major PLC vendors and some important end-users. Its main office is in the Netherlands but there are also supporting offices in Japan and the USA. By the summer of 1997, 11 products had achieved the PLCopen base level compliance for the Instruction List (IL) language. Other products are in the process of being evaluated for IL compliance. PLCopen is also developing test suites for compliance testing to the other IEC 1131-3 languages.

One aspect of the standard that may also reduce the portability of software between different systems is the provision for implementation dependent parameters. IEC 1131-3 defines over thirty parameters that can be defined by the PLC vendor. These include features such as the maximum number of array subscripts, the maximum number of function specifications and the maximum number of CASE selections. Clearly, for software to be truly portable, it must be built on systems that have very similar or identical IEC 1131-3 implementation dependent parameters.

The IEC 1131-3 standard has been criticised for defining too many optional features that may be implemented at the discretion of the PLC vendor. Having compliance levels that demand the implementation of a well defined set of features as proposed by PLCopen should help to circumvent this problem.

1.12 Inter-language portability

By providing various languages for expressing control software, IEC 1131-3 allows the software designer to choose the most suitable language for the problem in hand. Software elements such as function blocks can be programmed in any of the languages. The designer may select a particular language because it more clearly matches his or her mental realisation of the solution or simply because it is more familiar.

Although it was never a fundamental objective of the IEC group while developing the 1131-3 standard, it is generally possible to convert software elements expressed in one language into any of the other languages.

It is certainly possible to solve the same control problem using any of the languages although in some cases the algorithm may be slightly different.

However, some IEC language implementors have developed systems that will automatically convert between any two languages, say, from Ladder Diagram to Function Block Diagram, or Instruction List to Structured Text and so on. For example, a function block can be programmed using Function Block Diagram by one programmer, reviewed in Ladder Diagram format by a maintenance engineer and then documented in Structured Text for a code walk-through. This gives the exciting possibility that the person reviewing the software is able to choose a language that he or she finds the easiest to understand.

There are a few features in some languages that cannot be ported accurately to the other languages. Nevertheless the common set of features in the IEC languages is sufficiently large to allow a major proportion of software to be converted between any two languages. Problems that prohibit complete inter-language portability may be resolved in later revisions of the standard.

1.13 Graphical programming tools

The IEC 1131-3 standard is beginning to have a significant impact on the type of facilities being offered by programming stations used to design and program the next generation of PLCs and industrial control systems. The days when PLCs are programmed on character based screens using primitive graphics and obscure keyboard command sequences are numbered.

State-of-the-art IEC based graphical programming stations are now emerging that provide many of the facilities found with computer aided software engineering (CASE) systems, now commonly used for large scale mainstream software development. Many IEC programming systems offer graphical programming screens with multiple windows, mouse operation, pull-down menus, built-in hypertext help and so on. Such stations are also able to apply certain software verification checks as the software is being developed. For example, when two function blocks are connected, the system can check that the data types of all connected parameters are compatible.

We will start to see a new generation of PLC programming stations that provide many of the facilities needed for the complete PLC system design life-cycle - from initial system analysis, top-level design, coding, system building, testing and on-line diagnostics through to maintenance.

1.14 Future developments

A second edition of IEC 1131-3 is planned for publication in the middle of 1999. This will include the amendments outlined in Appendix 2. These generally enhance the usability of the standard rather than adding any major new features.

In the longer term, probably by the end of 2000, a major new revision of the IEC 1131-3 standard is being proposed. The new features that are likely to be considered will include a standard file exchange format to allow IEC 1131-3 designs to be ported between different vendors' programming stations, improvements for developing safety critical applications and compatibility with the IEC 1499 Function Block standard. IEC 1499 which is currently being developed by IEC working group SC65/WG6 will address the use of function blocks in distributed applications that encompass both PLC and fieldbus devices. It is the intention that IEC 1499 will provide a consistent function block methodology across PLC and DCS systems.

Summary

The main reasons for developing the IEC 1131-3 standard are:

- that there have been no consistent methods for programming PLCs;
- that conventional PLC ladder programming languages have a number of deficiencies that discourage the development of quality software.

Use of IEC 1131-3 languages is bringing a number of benefits, to both the PLC software developer and the end-user:

- IEC 1131-3 provides a variety of languages for solving different types of industrial control problem.
- It encourages the development of quality software through well structured design, the use of encapsulation and information hiding.

- Use of IEC 1131-3 will allow the same control software to be developed for different PLC products.
- Formalisation of re-usable software, especially function blocks, should result in improving the productivity of system designers through the provision of off-the-shelf solutions.
- IEC programming tools are emerging that offer support for the complete PLC software life-cycle.

Chapter 2

IEC 1131-3 concepts

Among the more important aspects of any programming system is the provision for software to be decomposed into small manageable parts. In this chapter we will review how complex programs can be broken down into different software elements which have clear and formal interfaces with each other. We will also see how the standard provides a number of different languages for developing software building blocks.

The reasons why the IEC software standard encourages programs to be constructed using either a top-down or bottom-up design approach will also be discussed.

You are advised to read this chapter first in order to gain an understanding of the main concepts in the IEC standard and how software is structured before proceeding to the chapters on specific IEC languages.

2.1 The IEC software model

In the course of developing the IEC 1131-3 standard, the IEC working group has had to consider the wider context of a PLC program. Every program has to exist in and interact with an environment. What is the nature of the boundary between a program and the rest of a system? Clearly, it is not possible to define the structure of a PLC program without a good understanding of its interfaces and interactions with the control system and external plant.

When a program is loaded into a PLC and is running, it requires the following types of interface to function:

- **I/O interfaces**: With every PLC system there is a requirement to read values coming from physical input channels that may be connected to sensors of various types such as micro-switches, pressure transducers, thermocouples etc. On the output side, there is a requirement to update values of physical output channels to drive a variety of actuators such as valve positioners, solenoids, servos, heaters and so on.

- **Communications interfaces**: Most PLC systems are required to exchange information with other devices and PLCs, drive operator displays and panels.
- **System interfaces**: There is also always a requirement for a system interface between the PLC program and PLC hardware. System services are required to ensure that the program can be initialised and run (executed) correctly as intended by the program designer. In practice, these services are generally provided by a combination of PLC hardware and embedded system software - sometimes referred to as system firmware.

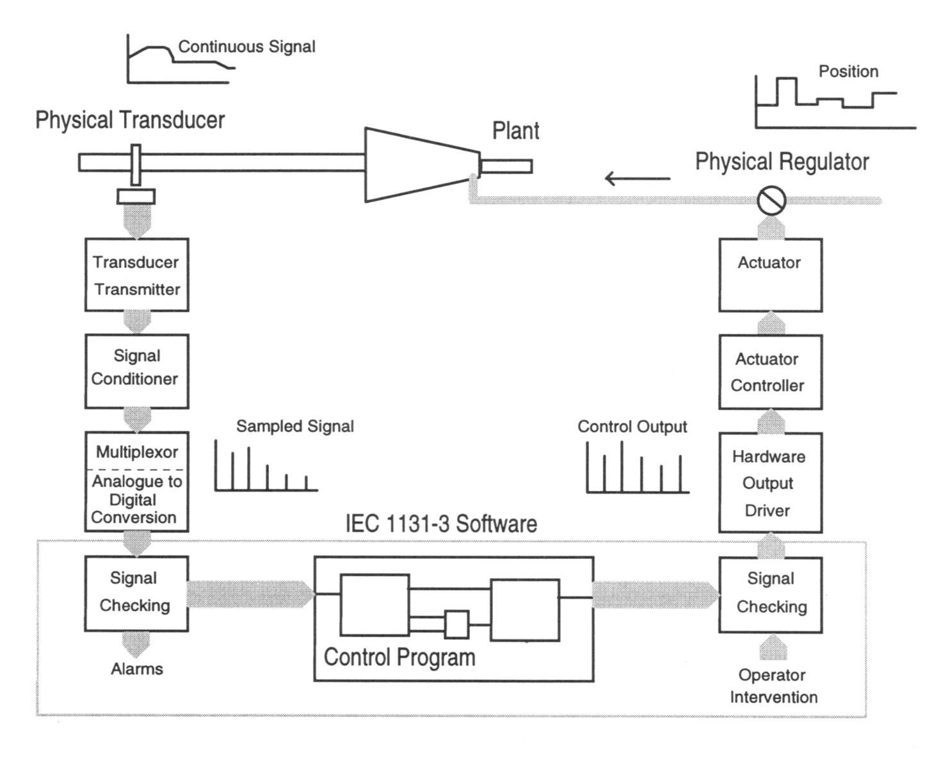

Figure 2.1 Using a PLC system for direct digital control

Traditionally many PLCs have been used in plant control applications where they provide interlock logic and alarm monitoring of purely digital signals. In such applications, inputs would typically come from digital sensors such as microswitches, and outputs would drive digital devices such as relays and indicators.

PLCs are now being used more frequently to provide complete plant control. In many cases that includes providing direct digital control (DDC) - sometimes referred to as closed loop control. Figure 2.1 depicts the principal components of a direct digital control system using a PLC. By converting continuous signals from a transducer into digital samples, a PLC control system can apply control algorithms such as a three term controller - proportional, integral, derivative (PID), to produce the desired output control signal that, in turn, can be used to position an actuator. The figure shows the speed of a steam turbine being controlled by a steam valve.

> Note: From general control theory, the 'sampling rule' sets the theoretical minimum rate at which a signal should be sampled in order to represent it completely as a sequence of numbers. In practice establishing the appropriate sampling rate for stable and responsive control may require detailed information on equipment response times, plant time constants etc. The theory behind this analysis is not in the scope of this book. It will always be necessary to calculate minimum acceptable sample rates for different control loops when designing a DDC system. However, with the IEC 1131-3 standard this is made somewhat easier because the designer now has the flexibility of being able to configure different parts of a control program to scan at different rates. [1]

The software and concepts that come within the scope of the IEC 1131-3 standard are shown outlined in Figure 2.1. The standard assumes that the values of external sensors and transducers are available from particular PLC memory locations. It also assumes that output values to drive actuators and indicators will be sent by updating certain PLC memory locations. The mapping of PLC memory to input/output (I/O) devices is not defined in the standard and will vary significantly from one manufacturer's PLC to another.

The main elements of the IEC software model needed to provide the PLC software environment are shown in Figure 2.2.

The model is layered - each layer hiding many of the features of the layers beneath it. We will now review the main features of this model to see how they are used together to build software for a PLC control application.

---

[1] *The provision of multiple tasks is an optional feature. Some PLC vendors may choose to provide a fixed set of tasks with only a limited selection of scan times.*

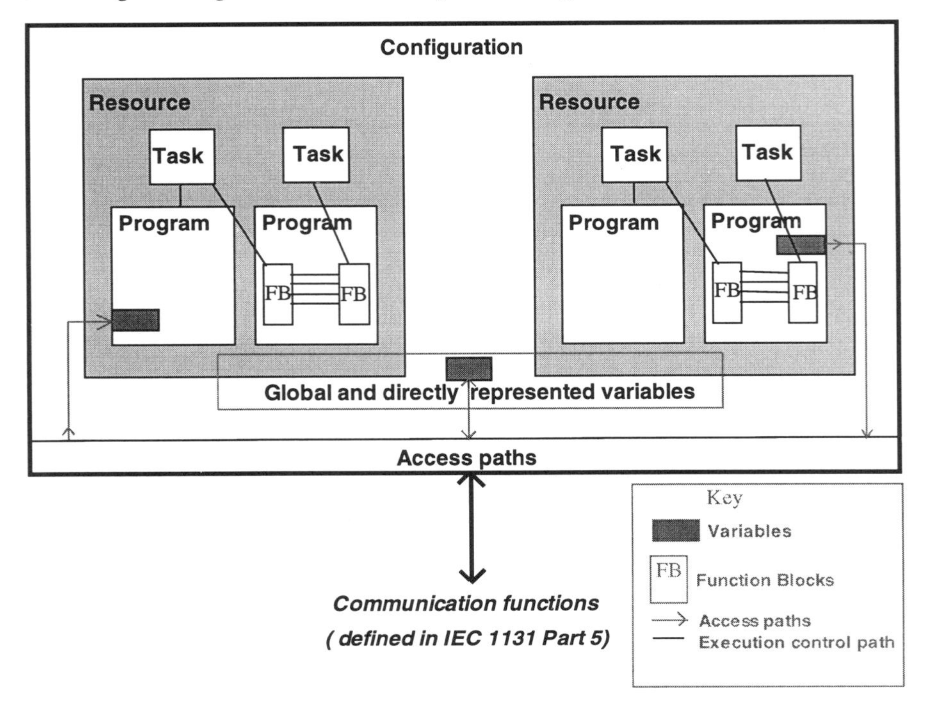

Figure 2.2 IEC 1131-3 software model

Configurations

At the highest level, the software for a particular control problem is contained in a **configuration**[2]. Existing at the outer layer of the entire PLC software model, a configuration is defined as a language element that corresponds to the *programmable controller system.*

Generally a configuration equates with the software required for one PLC. With large complex applications such as the automation of an entire production line, there may be a requirement for several PLCs to interact with each other, in which case the software for each PLC would be regarded as a separate configuration.

A configuration is able to communicate with other IEC configurations within different PLC systems via defined interfaces - all of which must be formally specified using standard language elements.

---

[2] *In IEC 1131-3 terminology, the word 'configuration' applies to the entire software that defines the behaviour of a PLC for a specific application. It should not be confused with the actual process of configuring a system.*

Resources

Within each configuration there are one or more **resources**. A resource provides support for all of the features needed for the execution of programs.

In software engineering terms, the resource may be regarded as the interface into a 'virtual machine' that is able to execute an IEC program. Its main function is to provide a support system to run programs.

An IEC program cannot function unless it is loaded into a resource. Normally a resource will exist within a PLC, but there is no reason why an IEC resource may not be provided by other systems. For example, a resource may be simulated within a personal computer to support PLC program testing. If it is possible to create an IEC resource, then it should be possible to execute IEC programs.

An interesting feature of a resource is that it is a subdivision of the software and may also reflect the physical structure of the PLC in which it resides. For example, the software for a PLC with multiple processor cards may be designed to have a resource on each card. The control of a resource in a configuration should be capable of independent operation - i.e. it should not be influenced by other resources.

One of the main functions of a resource is to provide an interface between a program and the physical I/O channels of the PLC.

The standard tersely defines a resource as:

> *'A language element corresponding to a "signal processing function" and its man-machine interface and sensor and actuator interface functions, if any as defined in IEC 1131-1'*

It is interesting that the standard allows a configuration to contain a number of resources and for each resource to be able to support more than one program. This gives the exciting possibility of a PLC being able to load, start and execute a number of totally independent programs.

> Note: In the course of designing the IEC standard, the working group considered an ideal software model that can support all future control requirements. Control systems and PLCs that can run multiple programs within multiple resources clearly may not be readily implemented using today's technology. With the development of yet lower cost and higher performance micro-processors and memory, this may clearly become a future possibility.

Programs

An IEC **program** can be built from a number of different software elements each of which may be written in any one of the different IEC languages. Typically a program consists of a number of interconnected **function blocks** that are able to exchange data through software connections. A program can read and write to I/O variables and communicate with other programs. The execution of different parts of a program, for example selected function blocks, may be controlled using **tasks**.

IEC 1131-3 defines a program as:

> *'A logical assembly of all programming elements and constructs necessary for the intended signal processing required for the control of a machine or process by a programmable controller system.'*

Tasks

A **task** can be configured to control a set of programs and/or function blocks either to execute periodically or upon the occurrence of a specified trigger. In this context, a single program or function block execution implies that all the software elements within the program are processed once.

In the IEC model, there are no hidden or implied mechanisms for execution of programs. In other words, a program or function block will remain completely dormant unless it is (a) assigned to a specific task and (b) the task is configured to execute periodically or when triggered by a specified variable changing state.

However, note that function blocks not explicitly associated with a particular task will always execute in the same task as their parent program.

Function blocks

The **function block** concept is one of the most important features within the IEC 1131-3 standard for supporting software hierarchical design. Using function blocks it is possible to create a program from smaller more manageable blocks. As any function block can also be programmed using other function blocks, it is possible to create a program that is well structured and truly hierarchical.

A function block has two main features: it defines *data* as a set of input and output parameters that can be used as software connections to other blocks and internal variables, and also has an *algorithm*, a piece of code that runs every time the function block is executed. The algorithm processes the current values on the input parameters and the values of internal variables to produce a new set of values for output parameters. In some cases the algorithm may create new values to update values of internal variables which may be local, i.e. declared within the function block, or global, i.e. declared at the program, resource or configuration level.

Because a function block is able to store values, it has a defined state. We will see that this is an important property that allows function blocks to be used to solve a wide range of control problems. For example, function blocks can be provided for closed loop control using PID or fuzzy-logic algorithms.

The standard specifies a range of standard function blocks for commonly required facilities including R-S bistables, counters, timers and real-time clocks. More importantly, IEC 1131-3 also allows the designer to produce new function blocks which can be built up from existing function blocks and other software elements.

Functions

The standard also specifies the use of **functions** that, understandably, are often confused with function blocks. Functions are not shown in the software model in Figure 2.1, but are discussed here to clarify what might otherwise be a misleading concept.

A function is a software element which when executed, with a particular set of input values, produces one primary result. The best examples are mathematical functions such as SIN() and COS() which produce one result for each particular input value. However, an amendment to the 1993 revision of IEC 1131-3 allows functions to also have VAR_IN_OUT parameters: see Section 3.11. This allows a function to modify the value of variables that are passed as VAR_IN_OUT parameters. For example, an array passed as a VAR_IN_OUT parameter could be modified within a function. A VAR_IN_OUT parameter is similar to a value passed by reference in languages such as PASCAL.

A large number of different types of function are specified in the standard for handling different types of data. Apart from trigonometric functions, there are functions for handling strings of characters, processing dates and time, selecting values and so on; Chapter 3 'Common elements' describes all the standard functions.

The difference between functions and function blocks

It is important to understand the difference between functions and function blocks. A function has no internal storage; it will always produce the same result for the same set of input values every time it executes. A function only has a single primary output.[3]

In contrast a function block may have numerous outputs. More significantly, a function block will not necessarily produce the same set of output values if it executes repeatedly with the same set of input values.

For example, consider a ramp function block that is used to increase the value of an output steadily from, say, a lower value to a higher target value at a given rate. Although the input values that define the ramp rate and final target value remain unchanged, every time the ramp function block is run, the output value will increase by a small amount until it reaches the desired target value. This is because a function block is able to store values in both output variables and internal storage.

Local and global variables

The standard permits variables to be declared within different software elements such as programs and function blocks. Variables can be given meaningful names and can contain different types of data. Variables are, by default, defined as **local** and can only be accessed within the containing software element. Local variables can be declared in a configuration, resource, program, function block or function.

Variables may also be defined as **global**. A global variable declared in a program can be accessed from within all software elements within the program, including within nested function blocks.

Similarly, global variables may also be defined at the resource level and accessible to all software elements within the resource or at configuration level where they are accessible by all software elements in the entire configuration.

Global variables typically provide a mechanism for transferring information and data between programs or between function blocks residing within different programs.

Directly represented variables

Memory locations within a PLC can be addressed directly using **directly represented variables.** The format of this type of variable may be familiar to anyone who has developed ladder programs and is in common use in many PLCs,

[3] *In the graphical languages, a function may have an additional input EN and output ENO for explicit execution control - see Section 3.28. Note also that the value of inputs that are declared as VAR_IN_OUT parameters may be modified within the function.*

for example %IW100, %Q75. Directly represented variables allow data to be written to and read from known memory locations within a PLC.

The use of such variables is restricted; they can only be declared and accessed within programs. The IEC has not permitted the use of directly represented variables within function blocks because their use would result in software that cannot be easily re-used.

It should be noted that the memory layout and, consequently, the addresses of directly represented variables will generally change between different applications. This means that software that uses a significant number of directly addressed variables will be difficult to re-use in other applications.

Since all programs within a configuration can generally access data stored in PLC memory, directly represented variables provide another way for programs to exchange information.

Access paths

So far we have reviewed the elements of the IEC software model that go together to form a configuration. We shall now look at the last feature of the model - **access paths** that provide facilities for transferring data and information between different IEC configurations. Each configuration may have a number of designated named variables which can be accessed by other remote configurations.

The standard assumes that there is some form of communications facility available to exchange data between different configurations but it is not specific about what form the communications should take. For example, configurations could communicate using a network based on Ethernet or Fieldbus or exchange data via a proprietary backplane bus. The lower layer protocols that might be used to provide communications facilities are beyond the scope of the IEC 1131-3 standard.

Access paths are provided by special variables declared using a construct known as VAR_ACCESS. Such variables may be read, or written to, by other remote configurations. In other words, the system designer can specify a set of variables that formally define the interface that a configuration presents to all other remote configurations.

Part 5 of the IEC 1131 standard defines communications function blocks that provide services to read and write to access variables in remote configurations. These are discussed in more detail in Chapter 12 'Communications'.

Control flow

Generally configurations and resources are responsible for the control and execution of elements within them. We will see that starting a configuration sets off a chain reaction that will result in starting all software elements that exist within it.

IEC 1131-3 states that external facilities (which are outside the scope of the standard) that are provided for the 'operator interface', for 'programming, testing and monitoring' or by the 'operating system' can be used to start and stop configurations and resources. Normally a PLC programming station will provide facilities to load and start various software elements including configurations and resources. The actual mechanisms to implement these facilities are not specified in the standard and will vary from one manufacturer's PLC to another.

The standard describes the behaviour for system start-up and shut-down.

Start-up

- When a configuration is started, all global variables are initialised and all resources are then started.
- When a resource is started, all variables within the resource are initialised, and all tasks are then enabled.
- Once tasks are enabled, all programs and function blocks assigned to the tasks will start to execute.

Shut-down

- When a configuration is stopped, all contained resources are stopped.
- When a resource is stopped, all tasks are disabled with the result that programs and function blocks cease to execute.

Note that programs do not always have control over the execution of all the function blocks that they contain. Certain function blocks may be assigned to tasks that cause them to execute periodically or in response to specific triggers. The time when a function block executes may not necessarily be synchronised with the time when its parent program executes.

Mapping the software model to real systems

Models are fine in theory but how can this software model be used in a real PLC system? Mapping the various parts of this software model onto a real system is very much an implementation issue.

Generally with small PLC systems with only one main processor (CPU) board, the model will typically degenerate to have one configuration, one resource and one program. The configuration contains all the software that defines the entire behaviour of the PLC system. The resource represents the capabilities of the processor board, while the program represents the signal processing part of the software - in a conventional PLC this would represent the ladder program. In a small PLC, the configuration will generally contain software to control a single item of plant or a machine.

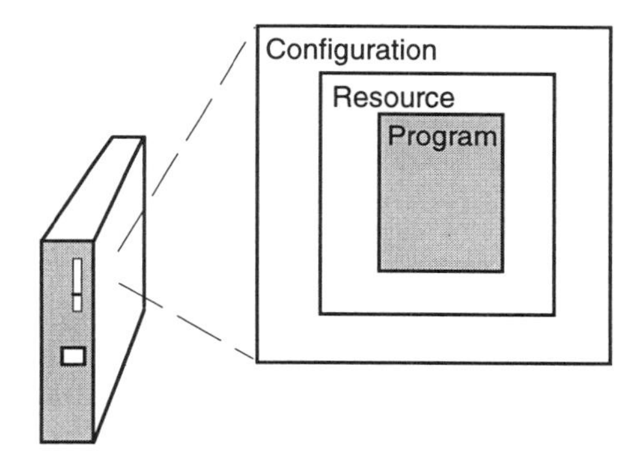

Figure 2.3 Small PLC with a single processor

With larger systems with multiple processors the mapping becomes more complex. The entire PLC may still be regarded as a single configuration. Each processor board will generally represent a resource. Depending on the memory and real-time multi-tasking capabilities of the processor, each resource may support one or more programs. An example of a multi-processor PLC is shown in Figure 2.4.

With large distributed systems, for example having many processors on a fast network using a fibre optic Fieldbus, various clusters of processing nodes may be regarded as one configuration or as a number of different configurations. How clusters of processors are aggregated into configurations is very much an implementation question. It will depend on how closely the various processors have to work together and how configurations are arranged to control the major items of plant.

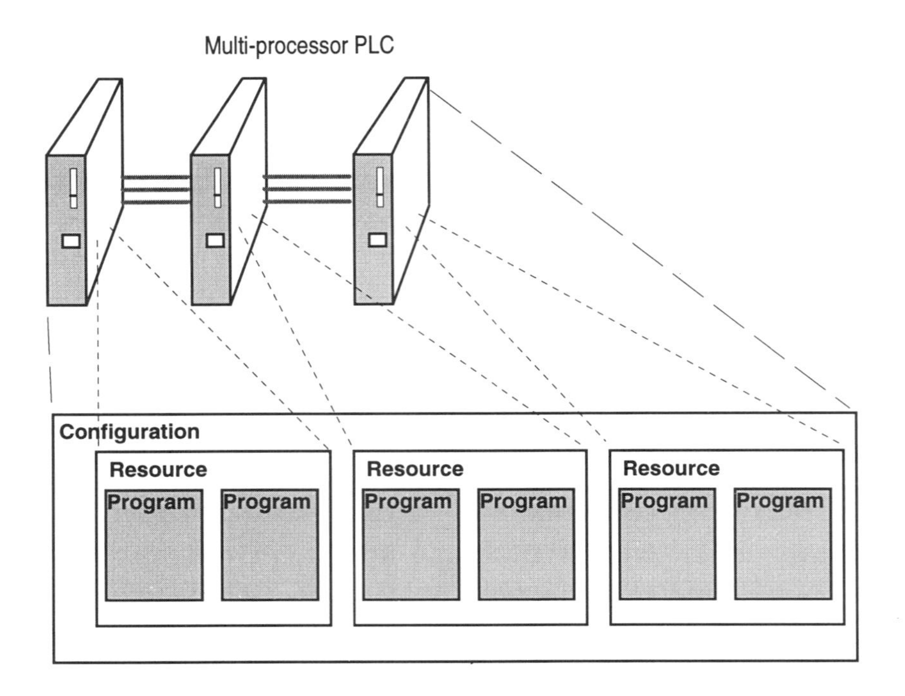

Figure 2.4 Multi-processor PLC

Figure 2.5 shows a number of processing nodes that are able to communicate via some form of high speed bus or network. Each node could be a single PLC processor board.

Configurations A and B are formed from resources that exist in processing nodes that are able to exchange data via a communications system. On condition that the communications system is able to allow clusters of resources to 'see' a common set of global and directly represented variables, configurations can be formed from resources that need not necessarily be physically near each other or even be in adjacent nodes in a network.

IEC 1131-3 provides language elements for defining and building configurations but does not describe mechanisms for managing distributed configurations. Clearly, this may become an important issue to be resolved in future work on the standard.

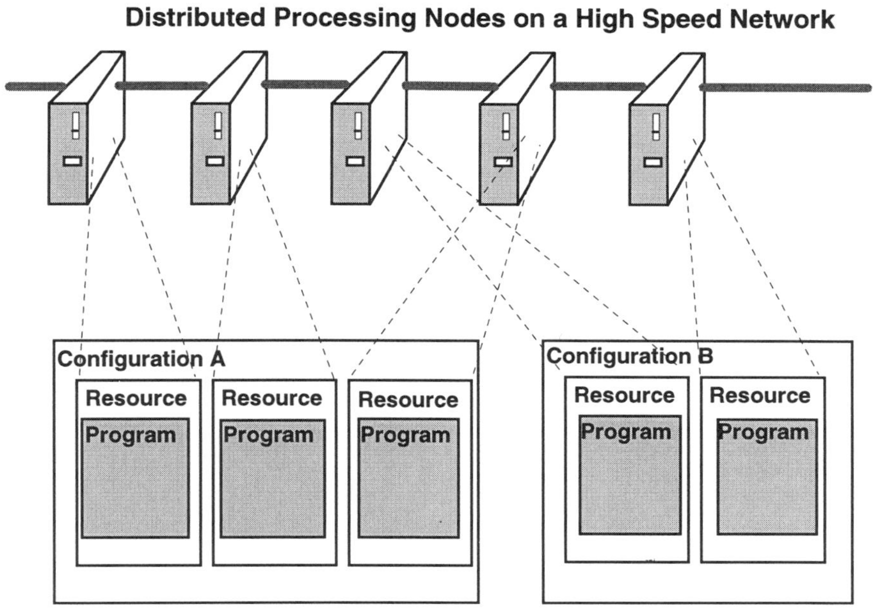

Distributed Processing Nodes on a High Speed Network

Configuration A

Resource Resource Resource

Program Program Program

Configuration B

Resource Resource

Program Program

Figure 2.5 Distributed processing

Mapping applications to IEC configurations

We have seen that IEC 1131-3 allows fairly complex configurations to be built from a number of resources and programs. The concept of an **application** is not defined in the IEC 1131-3 standard. We will discuss it here as it is an important part of the software model, particularly when we consider the design of PLC systems that are capable of controlling different units of plant or machinery. Clearly the end product of any configured PLC system is the solution of an industrial control problem.

A single solution, or more formally an **application,** must encompass all the control requirements to start-up, control and shut-down a machine or plant unit. The control will typically need to include the handling of physical sensors and actuators, interlocks and control loops, scheduling of events, the interaction with operator terminals and communications with remote devices. To achieve this an application may require the interaction of a number of IEC programs loaded into one or more resources.

Because IEC 1131-3 allows resources to be loaded and started independently, there is no reason why a large PLC system cannot handle a number of independent applications simultaneously. For example, a PLC system may be running several applications that independently control a number of reactor vessels, steam generators and compressors, while monitoring critical environmental conditions. Each application can be loaded and started independently if loaded into different resources.

The smallest application will always require one program. It is not feasible to have totally independent applications running within one program. This is because it is not possible to load, start and stop different parts of a program without disturbing the whole program.

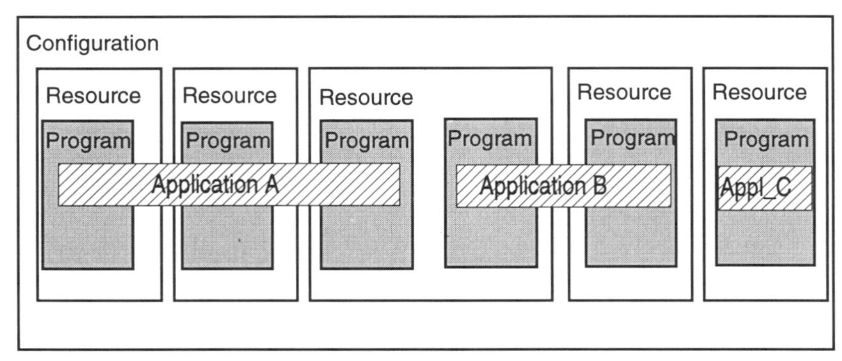

Figure 2.6 Mapping applications to resources

However, we will see that a program can itself handle parallel activities through the use of tasks and parallel sequences within Sequential Function Charts. Therefore, once initialised, and running, a program is able to control independently different machines or units of plant. However remember that, if it is necessary to modify the software in one part of the program, it may be necessary to stop the whole program in order to re-load a new version.

Program organisation units

The IEC standard describes programs, function blocks and functions as **program organisation units** or **POU**s. These software elements have the important property that their behaviour can be used repeatedly in different parts of an application. The behaviour and structure of a program organisation unit, such as a function block, are defined by a type declaration. Copies of function blocks made from a particular type are referred to as **function block instances**.

For example, a function block library may contain a Motor_Starter function block. A programmer may decide to create, say, six copies of this block within a program in order to control six different motors. Each copy will behave independently of the others, but will exhibit the same functionality, because each instance will use the same piece of code or algorithm. However, each copy will have its own private data area. Similarly, multiple instances of a particular type of program can be made within different resources. Multiple copies of particular functions can be made within IEC languages.

Program organisation units encourage software re-use right through from the macro level, with programs, through to re-use at a micro level, with function blocks and functions. The type definition of a program organisation unit can be regarded as a prototype from which instances with identical behaviour can be created. Therefore, each instance of a POU, for example each function block, can be assured to be error free provided that there are no errors in the associated type definition.

Table 2.1 Program organisation units

POU type	**Replicated as:**	**Comments**
Program type	Program instance	Provides macro level re-use of software, e.g. programs for boiler control, conveyor line management
Function block type	Function block instance	Allows complex through to simple control strategies and algorithms to be re-used, e.g. PID control, ramp, filter, lag
Function type	Function	Used for commonly required data manipulation functions, e.g. COS, SIN, MAX, CONCAT

Unlike other high level languages such as PASCAL, IEC 1131-3 specifically forbids the use of recursion within POUs. For example, a function cannot call itself within the code of its own declaration. This is clearly because recursive software is difficult to test and its real-time performance is not predictable.

Programs and function blocks can be programmed using any of the IEC lanaguages, i.e. Structured Text, Ladder Diagram, Function Block Diagram, Instruction List or Sequential Function Chart. As functions have no retained state information they can be programmed in all languages except Sequential Function Chart.

Hierarchical design

One of the great strengths of the IEC 1131-3 standard is the strong emphasis on hierarchical design. We have seen that a program can be defined as a network of function blocks and functions. Each function block is a copy or an instance of a type definition.

A function block type definition may in turn be defined using instances of other function block types and so on. This allows complex programs to be broken down first into large function blocks representing major areas of functionality that in turn can be broken down into smaller blocks representing more focused areas of functionality. Ultimately, at the bottom level function blocks can either be defined using functions and textual language statements or be supplied from standard libraries.

Most IEC 1131-3 based systems will provide libraries containing a wide variety of different function block types, e.g. PID, ramp, filter, lag. With some systems it will also be possible to create project specific function block types, for example for a proprietary control strategy for a particular type of plant. The same applies to functions which can also be defined hierarchically. The program designer is able to define new functions using language statements operators and other functions - some may be taken from standard libraries, e.g. COS(), LOG().

The number of layers of hierarchical decomposition may vary between different proprietary PLC systems. However, some ability for hierarchical decomposition is essential to be in keeping with the spirit of IEC 1131-3.

In Figure 2.7 we can see an example of a program A1 that is constructed from a network of connected function block instances. Function block instance R1 has behaviour defined by its type definition. Within the definition for function block type R there is an instance X1 of function block type X.

Note: IEC programming stations will usually provide facilities to allow the designer to 'push-down' on any function block to reveal its type definition, so that exploring the design hierarchy is generally fairly straightforward.

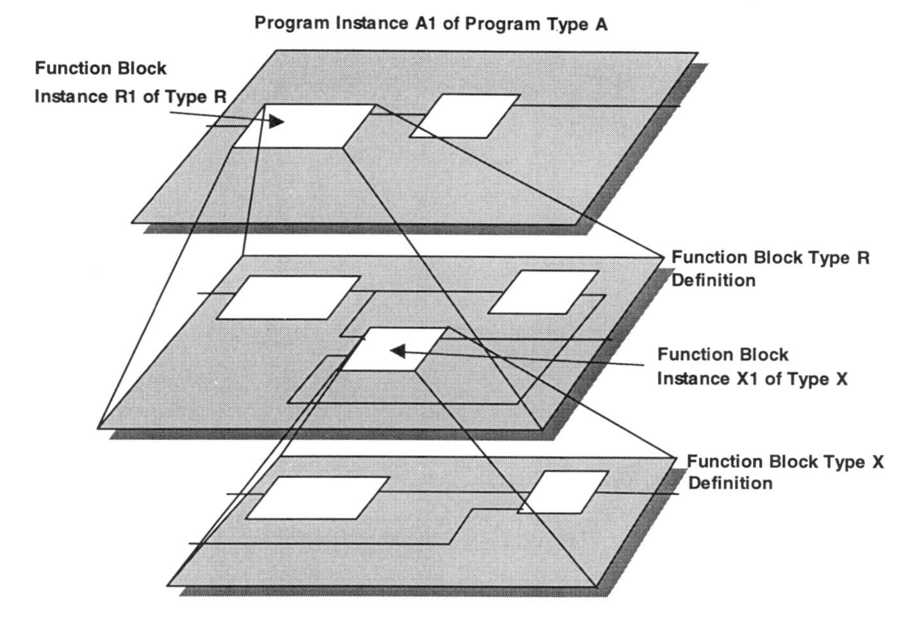

Figure 2.7 Hierarchical software decomposition

2.2 Communications model

Internal communications

Various mechanisms are provided to enable information to be exchanged between program organisation units, i.e. between programs, function blocks and functions. Within a program, function blocks and functions can be interconnected graphically to form a network. For example, values of variables can be passed from function block outputs to one or more function block inputs. Functions can also be used to modify values or produce derived values to be passed on to further functions or function blocks.

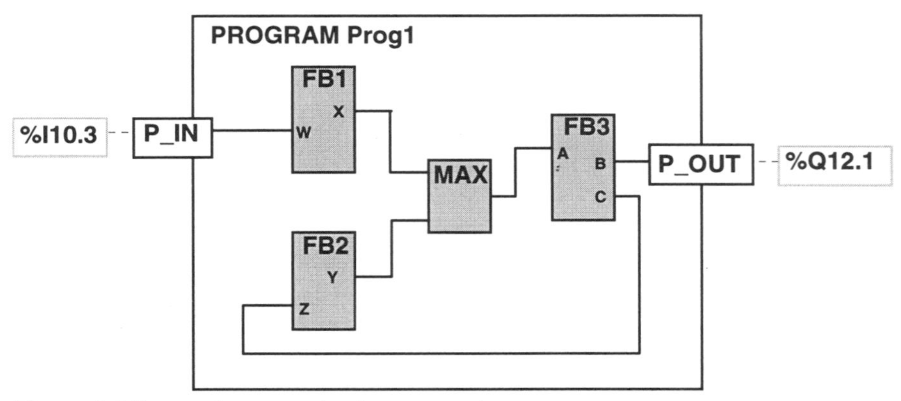

Figure 2.8 Internal communications example

At the program level, program inputs can be used to pass data into internal functions and function blocks from external variables. Generally program inputs will be connected to variables that are associated with data coming from physical devices such as PLC input sensors. Similarly program outputs which are driven from outputs of function blocks and functions within the program are generally connected to variables associated with physical output devices.

Figure 2.8 shows some examples of internal communication paths between program organisation units. Program Prog1 is defined using three function blocks FB1, FB2 and FB3 and a function MAX. The input parameter A of FB3 is produced by taking the maximum value of parameters X and Y from function blocks FB1 and FB2 respectively, produced by the MAX function. Parameter C of FB3 forms a feedback path passing data back to input Z of FB2. Program input parameter P_IN allows data to be passed into the program, in this case to input W of FB1. Output parameter P_OUT allows data to be passed out from the output B of function block FB3.

In this example program input P_IN is configured to receive the current value of directly represented variable %I10.3, which could be the memory location of the value from a particular input sensor. Program output P_OUT is connected to directly represented variable %Q12.1 that could, for example, be the memory location for updating the value of a particular output device.

The Function Block Diagram language allows such data paths to be depicted graphically. However, data can also be exchanged between POUs using Ladder Diagram language and the textual languages, Structured Text and Instruction List.

Using globals

Global variables can also be used to exchange data between function blocks and other POUs. Global variables can be defined at the configuration, resource or program level. Globals are accessible to all program organisation units contained within the parent entity in which they are declared. For example, a global variable declared within a program is accessible to function blocks and functions within that program. Global variables defined at the configuration level can be accessed by all programs and function blocks within all resources contained in the configuration.

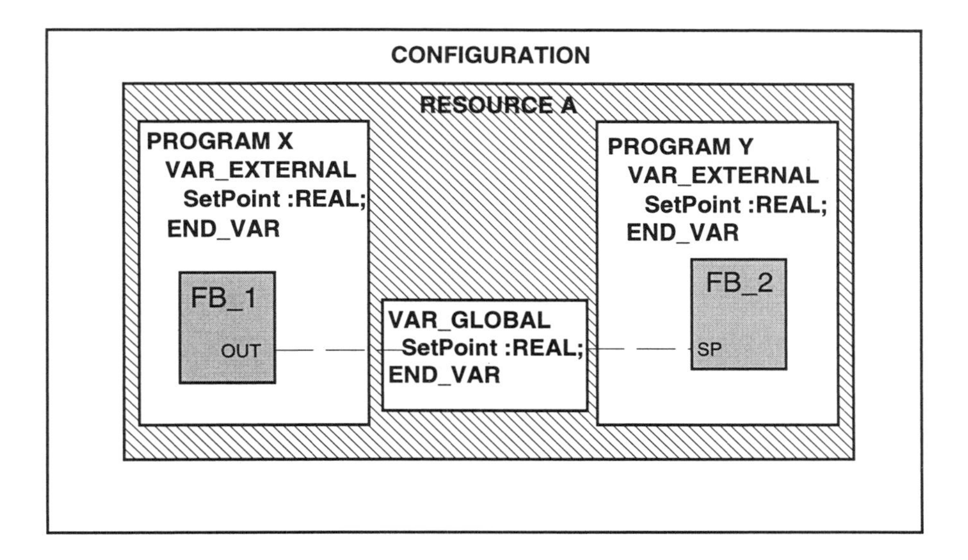

Figure 2.9 Internal communications using globals

Figure 2.9 depicts the use of a global variable to exchange data between function blocks FB_1 and FB_2 in different programs. The global variable SetPoint is declared within the Resource A. By declaring the SetPoint as an external variable in both programs it is accessible by function blocks in both programs. In this example, the output OUT of function block FB_1 is used to update the global SetPoint. Function block FB_2 is able to read the current value of SetPoint on its input SP.

Globals are a very flexible method of exchanging data between different programs and function blocks. However because the data can be written and read from global variables at various points within a program or function block, programs developed using numerous global variables can be difficult to understand. Data flows created by directly connecting inputs and outputs of programs and function blocks can be depicted graphically and result in more understandable and therefore more maintainable software.

External communications

Part 3 of the IEC 1131 standard only partly defines the external communications facilities required to exchange data between programs in different configurations. Nevertheless, Part 5 of IEC 1131 defines a family of communications function blocks that can be used to exchange data over a network. The IEC 1131-5 function blocks were primarily designed to operate over a MAP based network using the Manufacturing Messaging Specification (MMS). However some manufacturers may choose to provide similar function blocks to run over their own proprietary networks.

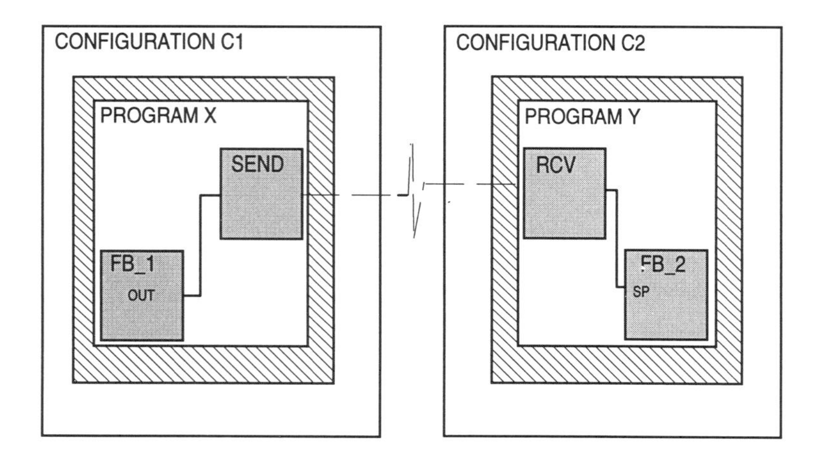

Figure 2.10 Using communications function blocks

An example of using the IEC 1131-5 communications function blocks is shown in Figure 2.10. The figure depicts the transfer of data from an output of a function block FB_1 in configuration C1 to the input of a function block FB_2 in a remote configuration C2. A communications function block of type SEND is used to provide network addressing facilities and control the data transfer to a communications function block in a remote configuration, in this case to a function block of type RCV, i.e. for receiving data.

The communications function blocks hide the complexity of handling network data transfers and significantly simplify the design of networked applications.

Using access paths

Access variables provide a method in IEC 1131-3 for making particular variables accessible from remote devices.

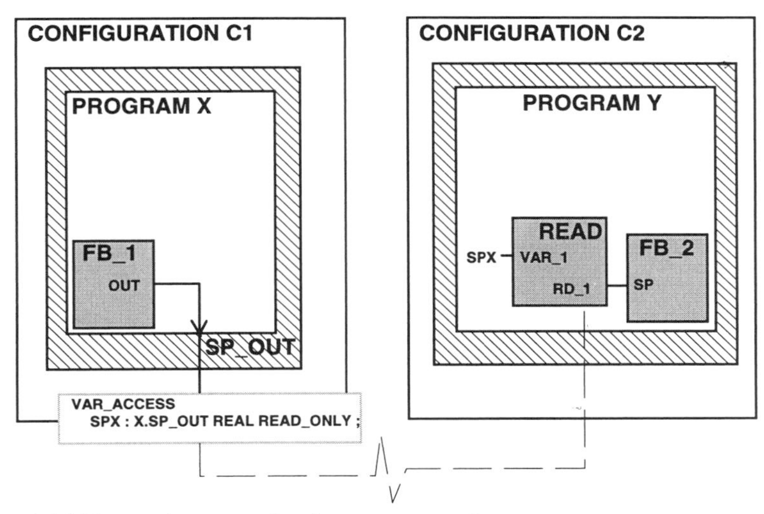

Figure 2.11 External communications access paths

Communications function blocks defined in part 5 of the IEC 1131 standard which is concerned with PLC communications can also be used to read and write to ACCESS variables in a remote configuration. Figure 2.11 depicts the use of a READ function block to update input SP of function block FB_2 from the value of the variable read from a remote configuration C1.

The value of this variable originates from the output OUT of function block FB_1. The value of FB1.OUT is assigned to program X output parameter SP_OUT. An ACCESS variable named SPX provides external 'read only' access across the communications system to the value of program output X.SP_OUT. The READ function block declared within program Y is only able to read the value of X.SP_OUT using the alias SPX from within the configuration C1.

The use of communications function blocks is discussed in more detail in Chapter 12.

Access variables may also be read and written to by other devices on a communications system, such as cell controllers and SCADA work stations.

Summary

We have seen that the IEC 1131-3 standard has adopted a well defined and formal model for the design of software within a PLC system. The model has been developed to cater for typical industrial applications of PLCs, including use in direct digital control systems.

The IEC 1131-3 software model has a number of strengths:

- It is flexible and can be applied to a wide range of different PLC architectures.
- It is suitable for both small scale systems and large distributed systems.
- It encourages hierarchical design decomposition.
- Software can be designed for re-use as program organisation units (POUs), i.e. as programs, function blocks and functions.
- It provides facilities for exchanging information via communications networks.

Chapter 3

Common elements

In this chapter we introduce many of the common programming elements that are used with all of the IEC languages. An understanding of how elements such as variables and data types are used is necessary before it is possible to go on to develop programs using Structured Text or any of the other languages.

We will see that IEC 1131-3 provides an extensive range of common elements that can be used both with the textual languages Structured Text (ST) and Instruction List (IL), and with the graphical languages Function Block Diagram (FBD) and Ladder Diagram (LD).

Note: The languages ST, IL, FBD and LD and the common elements discussed in this chapter can be used within action blocks and transitions to construct Sequential Function Charts (SFCs) for describing time- and event-driven behaviour - for further information see Chapter 8, 'Sequential Function Chart'.

In this chapter we will include descriptions of elements for:

- Naming software elements using identifiers, such as for variables;
- Declaring variables using standard types of data;
- Declaring variables from derived data types, structures and arrays of data;
- Establishing the default initial values for certain types of data;
- Ensuring that variables are initialised correctly.

We will then proceed to describe the main software building blocks of the language, for which the IEC has the generic name program organisation units or POUs. These are specifically functions, function blocks and programs.

Finally we will discuss how complete PLC system configurations can be declared using resources, tasks and programs.

This chapter covers all common programming elements with the exception of steps, transitions and actions which are discussed in Chapter 8, 'Sequential Function Chart'.

3.1 Common programming elements

The IEC standard defines a large number of common features and elements that apply to all the programming languages. The internal behaviour of program organisation units, i.e. programs, function blocks and functions, can be described using any one of five different languages. However, irrespective of the language used, the variables and data types handled by all POUs are described using the same common programming elements.

For example, input and output variables of a function block will be described in the same terms whether the function block is programmed using Ladder Diagram, Function Block Diagram or Instruction List.

3.2 Character set

To ensure that IEC programs can be readily ported to different systems, all textual information should use a restricted set of letters, digits and characters. The standard requires that only characters from standard ISO 646 "Basic code table" are used. In practice, this character set is fairly generous and uses characters that are in everyday use on a PC; there are no unusual symbols.

The standard provides alternatives where there are conflicts between pound sign '£' and number or hash sign '#'. Where a national character set does not have a vertical bar character '|' for character based graphics, an exclamation mark can be used.

The use of special national characters, such as Æ, is not permitted but such letters could be included as a national extension to the standard.

Optionally lower case letters can be used but they are not significant in language elements; for example, Heater1 and HEATER1 are treated as identical. However lower case letters are still significant if used to define printable strings such as 'Load Job 1a'. Upper and lower case letters can be used within comments.

Note: Identifiers are case insensitive but language keywords are case sensitive and should always be in uppercase.

3.3 Identifiers

Identifiers are used for naming different elements within the IEC languages, for example for naming variables, new data types, function blocks and programs.

An identifier can be any string of letters, digits and underlines provided that:

(*a*) The first character is not a digit;

(*b*) There are not two or more underline characters together.

The standard provides options for supporting identifiers with lower case letters, and identifiers with a leading underline character. There can be no embedded spaces.

The following are acceptable identifiers when all options permitted by the standard are supported:

```
W123_PV     W12_3PV     aTemp1     _PROG1     ✓
```

The following are illegal:

```
W123__PV    W12 3PV     1Tempa     Q%TY12     ✗
```

The standard states that at least the first six characters should be tested for uniqueness. The following two identifiers may therefore be regarded as identical on some systems:

```
A12345_XY A12345_GG
```

Consequently to avoid the possibility of ambiguous identifiers, when developing software that may be ported to different PLC systems, always ensure that the first six characters are unique.

3.4 Language keywords

Keywords are special words[1] that are used within the IEC languages to define different constructs or the start and end of particular software elements. For example, a function type definition is framed with keywords FUNCTION and END_FUNCTION.

A full list of keywords used by the IEC 1131-3 standard is given in Appendix 1. As a general rule, the use of identifiers that may be confused with language keywords should be avoided even though many language compilers may be able to distinguish between them from their position in the program. Avoid identifiers such as:

---

[1] *Keywords should be in upper case characters. However, an amendment to the 1993 standard for the second edition of IEC 1131-3 proposes that mixed case keywords should be allowed.*

```
TYPE TRUE PROGRAM TASK RETURN STEP FUNCTION
```

The identifiers of IEC 1131-3 standard function blocks and functions as described in Appendix 1 should be regarded as reserved, such as:

```
TON     RS     SIN     COS
```

3.5 Comments

Various length comments from short to multi-line can be inserted in any of the IEC languages. With the exception of Instruction List language where there are some restrictions, comments can be placed wherever it is acceptable to insert one or more spaces.[2] Comments are framed by the characters (* .... *).

Examples are:

```
(* Activate Pump *)
(**********************************)
(* Main turbine interlock logic *)
(**********************************)
```

3.6 Data types

As PLCs are now being used in an ever widening range of industrial applications, IEC 1131-3 recognises the need for handling a variety of different types of data. The standard provides a comprehensive range of standard elementary data types designed for dealing with values of typical industrial variables.

The range[3] includes floating point (REAL)[4] numbers for arithmetic computations, integers for counts and identities, booleans for logic operations, times and dates for timing and managing batch systems, strings for holding textual information and bits, bytes and words for low level device operations.

The format of literals for each data type is also given in the following descriptions. A literal is a constant that defines how a given value for a particular data type is represented; e.g. integer literals are 12, 342, 0. Where the data type of a literal may be ambiguous, the data type can be prefixed, e.g. INT#12, LINT#342.

---

[2] *Comments nested within comments will be regarded as an error.*

[3] *There is no requirement for a PLC product to support all data types defined in IEC 1131-3.*

[4] *Where an informal data type is given in the text, the formal IEC data type may sometimes be given in parentheses, e.g. floating point (REAL).*

Integer

There are a number of different integer types defined. Integers are used for holding values such as counts and identities, i.e. whole numbers. The 'unsigned' set of integers should be used where there is no requirement to hold negative values and where the increased positive range can be exploited. The type of integer used for a particular variable will depend on the range of values that need to be stored in the variable.

Table 3.1 Integer data types

IEC data type	Description	Bits	Range
SINT	Short integer	8	- 128 to + 127
INT	Integer	16	-32768 to 32767
DINT	Double integer	32	-2^{31} to $+2^{31}-1$
LINT	Long integer	64	-2^{63} to $+2^{63}-1$
USINT	Unsigned short integer	8	0 to 255
UINT	Unsigned integer	16	0 to $2^{16}-1$
UDINT	Unsigned double integer	32	0 to $2^{32}-1$
ULINT	Unsigned long integer	64	0 to $2^{64}-1$

For example, if a variable is going to be used to count the number of items on a conveyor belt, and the number can never exceed 100, then an integer of type SINT would be suitable.

However, if an integer variable was going to be used to count pulses from a shaft encoder where large positive and negative values can be expected, then a long integer LINT would be recommended.

Integer literals

These can be represented using decimal values, e.g. -123 0 +463 23_123

As binary values, i.e. base 2 values, e.g. 2#1111_1111 (255 decimal)
2#0000_1000 (8 decimal)

As octal, i.e. base 8 values, e.g. 8#377 (255 decimal)
8#020 (16 decimal)

As hexadecimal, i.e. base 16 values, e.g. 16#FF (255 decimal)
16#A0 (160 decimal)

Note: The standard allows the underline character '_' to be inserted into numeric literals to aid readability; otherwise it has no significance.

Floating point (REAL)

Table 3.2 Floating point (REAL) data types

Data type	Description	Bits	Range	
REAL	Real numbers	32	$\pm 10^{\pm 38}$	Note 1
LREAL	Long real numbers	64	$\pm 10^{\pm 308}$	Note 2

Note 1: REAL values have a precision of 1 part in 2^{23}
Note 2: LREAL values have a precision of 1 part in 2^{52}

The data types REAL and LREAL are used to store a wide range of floating point values, i.e. analogue values that can be either positive or negative, and can be from very large to very small fractional values, e.g. 342.0032, -0.000123, +1345000.0. The LREAL variable can store values of a much larger range and higher precision than REAL.

The format is defined by standard IEC 559; this is identical to the commonly used IEEE floating point number format.

Typical uses of REAL and LREAL data types are:

- Holding the values from analogue inputs, such as from pressure transducers, thermocouples, strain gauges, tachometers;
- Within algorithms used for closed loop control such as PID;
- Driving analogue outputs such as valve position actuators.

REAL literals

These can be represented using normal decimals and can be distinguished from integer literals by the presence of a decimal point. For large and very small values, the exponential notation can be used giving the power of ten by which the number can be multiplied; either upper case 'E' or lower case 'e' can be used to denote the ten's exponential value, e.g.

10.123, +12_123.21, -0.001298, -1.65E-10, 0.9E20, 0.3276e+14

Duration (TIME)

Data type	Description	Bits	Usage
TIME	Time duration	Note 3	Storing the duration of time after an event

Note 3: The length and precision of this data type are implementation dependent. The TIME data type is used to store durations, i.e. in days, hours, minutes, seconds and milliseconds.

Typical uses of this data type are:

- Defining the duration of a process phase; for example, heat treatment of materials, timing how long a moulding should be at a given temperature.
- Defining time-outs for particular events to occur; for example, timing when a valve should shut and raising an error condition if the close condition does not occur within a specified duration.

TIME literals

There are two forms of time literals provided, short form and long form. The short form is concise and fairly readable but where improved readability is required, the long form can be used.

In both forms the following letters are used:

d = days, h = hours, m = minutes, s = seconds, ms = milliseconds

Short form examples:

T#12d3h3s defines 12 days, 3 hours and 3 seconds
T#3s56ms defines 3 seconds and 56 milliseconds

Long form examples:

TIME#6d_10m defines 6 days and 10 minutes
TIME#16d_5h_3m_4s defines 16 days, 5 hours, 3 minutes and 4 seconds

The last field of a time literal can also be given in a decimal format.

Examples:

T#12d3.5h defines 12 days, 3.5 hours
T#10.12s defines 10.12 seconds

The most significant unit in the time duration literal can 'overflow'. For example the following are accepted:

T#61m5s - equivalent to T#1h1m5s
TIME#25h_3m - equivalent to TIME#1d_1h_3m

Dates and times of day

Table 3.3 Dates and times of day data types

Data type	Description	Bits	Usage
DATE	Calendar date	Note 4	Storing calendar dates
TIME_OF_DAY or TOD	Time of day	Note 4	Storing times of the day, i.e. real time clock
DATE_AND_TIME or DT	Date and time of day	Note 4	Storing the date and the time of day

Note 4: The length of this data type is implementation dependent.

These date and time data types have a wide range of uses particularly in batch oriented applications.

Typical usage of dates and times of day are:

- Recording the date and time of events and alarm conditions for fault diagnosis and process audit purposes.
- Controlling when certain events occur during the day, the week or even the year. For example, a reactor could be pre-heated automatically early every Monday morning ready for the weekly batch run.
- Recording when power fails and power resumes to calculate the system down-time.

Time of day and date literals

There are two forms of time of day and date literals, short form and long form. The short form is concise and fairly readable but where improved readability is required, the long form can be used.

Table 3.4 Dates and times of day literal prefixes

Data type	Short form	Long form
DATE	D#	DATE#
TIME_OF_DAY	TOD#	TIME_OF_DAY#
DATE_AND_TIME	DT#	DATE_AND_TIME#

Examples of DATE literals are:

D#1994-06-10 - 10th June 1994
d#1995-01-13 - 13th January 1995
DATE#2000-10-15 - 15th October 2000

Examples of TIME_OF_DAY literals are:

TOD#10:10:30 - 10 o'clock, 10 minutes and 30 seconds
TOD#23:59:59 - 1 second to midnight
TIME_OF_DAY#05:00:00.56 - 0.56 seconds past 5

Examples of DATE_AND_TIME literals are:

DT#1993-06-12-15:36:55.40 - 12th June 1993, at 15 hours, 36 minutes and 55.4 seconds
DATE_AND_TIME#1995-02-01-12:00:00 - midday on 1st February 1995

Strings

Data type	Description	Bits	Usage
STRING	Character strings	Note 5	Storing textual information

Note 5: The method used to store this data type is implementation dependent. Character strings can be used to hold textual information which may consist of both printable and non-printable characters.

Typical uses are:

- Holding batch identities, e.g. 'JOB_X32A3';
- Messages for operator displays, e.g. 'Starting Vessel Purge';
- Messages to be sent via communications drivers to other devices.

STRING literals

A string literal defines a number of printable and non-printable characters. All string literals should be framed using a single quote character ('). Non-printable characters can be inserted by prefixing the hexadecimal value of the character (given as two hexadecimal digits) with a dollar $. There are also a number of reserved letters which can be used after the dollar to denote commonly used control characters. Techniques for embedding control characters are shown in the following table:

Table 3.5 Embedding control characters within strings

Code	Interpretation
$$	Single dollar sign
$'	Single quote character
$L or $l	A line feed character
$N or $n	A new line character
$P or $p	Form feed, new page
$R or $r	Carriage return character
$T or $t	Tabulation, i.e. tab character

Examples of string literals are:

'Batch number AX45_65'
'End of report $N' — embedded line feed
'$01$02$10' — 3 characters with decimal codes 1, 2 and 16
'' — an empty or null string

Bit strings

Table 3.6 Bit string data types

Data type	Description	Bits	Usage
BOOL	Bit string of 1 bit	1	Digital, logical states
BYTE	Bit string of 8 bits	8	Binary information
WORD	Bit string of 16 bits	16	" "
DWORD	Bit string of 32 bits	32	" "
LWORD	Bit string of 64 bits	64	" "

A range of bit string variables is provided for storing binary information. These data types may be required when setting various status bits to be sent to remote devices and instruments or when handling low level interfaces to the PLC hardware.

Boolean

The Boolean data type is used to define the state of boolean variables, such as those associated with digital inputs and outputs.

IEC data type	Description
BOOL	Has two states, FALSE equivalent to 0, and TRUE equivalent to 1.

A boolean variable can be assigned logical values FALSE or TRUE. Alternatively the logical values 0 and 1 can be used.

Boolean variables are typically used in interlock or combinatorial logic and for holding the states of devices, for example whether a pump is 'ON' or 'OFF'. Single bits within the PLC memory are normally defined as variables of type BOOL.

Boolean literals
FALSE and TRUE can be used to define the literal values of booleans and are treated as reserved language keywords.

FALSE	0
TRUE	1

Bit string literals
Bit string literals are used to define the contents of bit strings having multiple bits. The same representation as defined for integers, e.g. 146, 2#1101, 16#FFAC, can be used.

3.7 Use of generic data types

The elementary data types that have similar properties are organised in a hierarchy as shown in Figure 3.1.

Generic data types that start with the prefix ANY_ can be used for describing variables in function and function blocks where there is support for overloaded inputs and outputs. Overloaded refers to the ability of a variable to be used for different types of data.

For example, a function AVE() for calculating the average value of two inputs could have inputs and an output all of data type ANY_NUM. This would imply that the function can be used with any data type from the hierarchy that comes under the generic data type ANY_NUM. In other words AVE could be used with different types of numeric data, such as decimal values (REAL), or integers (INT, UINT). A number of the IEC standard functions such as MAX() are described using generic data types indicating that the same function can be used for a variety of different types of data.

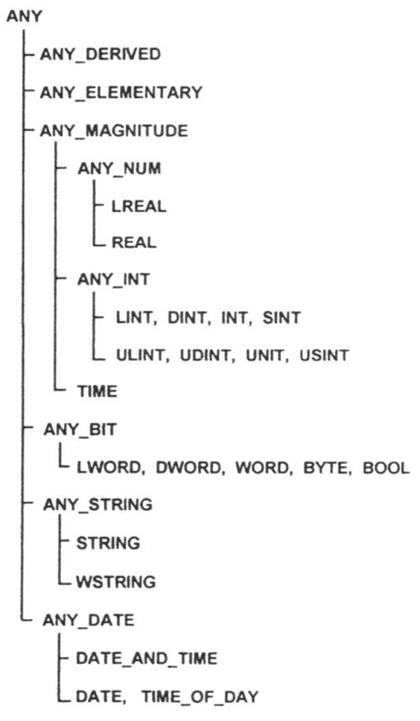

Figure 3.1 Hierarchy of elementary data types

Note 1: A PLC manufacturer should list the functions that support data type overloading. User-defined functions cannot be overloaded, i.e. cannot be defined using the ANY_... data types. Generic data types ANY_... are not IEC keywords.

Note 2: The standard requires that all inputs and outputs for an overloaded function must be of the same data type. This should always be checked by the language compiler.

Note 3: All subranges and array subscripts are data types ANY_INT.

Note 4: The generic type of a directly derived type will be the same as the generic type of the elementary type from which it was derived.

3.8 Initial values of elementary data types

An important principle in the IEC standard is that all variables have a known initial value. By default all variables are initialised to zero or in the case of strings to an empty null string. Dates default to the value of D#0001-01-01.

We will see that these default initial values can be overridden when a variable of a particular data type is declared.

3.9 Derived data types

To assist program readability and ensure that data are used consistently, new data types can be defined from the elementary data types. A new data type is declared by framing the definition with TYPE and END_TYPE.

For example, all values for pressures could be held in a derived data type called PRESSURE which is based on the REAL data type. The type can be declared as:

```
TYPE
     PRESSURE : REAL;
END_TYPE;
```

Structured data types

New composite data types can be derived by defining a structure from existing data types. A structure is declared by framing the definition with STRUCT and END_STRUCT. Each element of the composite data can be given a meaningful name.

For example, the data from a pressure sensor may consist of the following information:

(*a*) The current pressure as an analogue value;

(*b*) The status of the device, i.e. operating or faulty;

(*c*) The calibration date;

(*d*) The maximum safe operating value;

(*e*) The number of alarms in the current operating period.

This can all be defined in a single structure as a new data type PRESSURE_SENSOR:

```
TYPE PRESSURE_SENSOR:
  STRUCT
    INPUT:        PRESSURE;
    STATUS:       BOOL;
    CALIBRATION:  DATE;
    HIGH_LIMIT:   REAL;
    ALARM_COUNT:  INT;
  END_STRUCT;
END_TYPE
```

Enumerated data types

Enumerated data types allow different states of a value to be given different names.

For example, all devices within a particular system may have a number of different operational modes. A data type suitable for variables holding device modes would be:

```
TYPE DEVICE_MODE:
   (INITIALISING,RUNNING,STANDBY,FAULTY);
END_TYPE
```

In this case, variables of type DEVICE_MODE can only be assigned values from the list of named states, i.e. as given in the enumerated list[5].

To avoid ambiguities when using enumerated variables sharing identical enumerated literals, the data type of a particular enumeration literal can be specified using a prefix in the form '<data type>#'. Examples are:

```
TYPE
  VALVE_MODE: (OPEN, SHUT, FAULT);
  PUMP_MODE: (RUNNING,OFF,FAULT);
END_TYPE;
...
IF AX100 = PUMP_MODE#FAULT  THEN
       XV23 = VALVE_MODE#OPEN;
```

[5] *The standard does not permit a value to be associated with a named state or for enumerated values to be used to index arrays.*

Sub-range data types

There is often a requirement to restrict the range of values that can be assigned to integer types of variable. A sub-range can be specified that limits the value range.

For example, variables used to store voltages for controlling a set of DC motors may be restricted to values +12 volts and -6 volts. A suitable data type for MOTOR_VOLTS would be:

```
TYPE
   MOTOR_VOLTS: INT(-6..+12);
END_TYPE
```

It is expected that the programming station or compiler will provide range checks that ensure the variables of sub-range data types are only assigned values inside the range.

Array data types

In many PLC applications there is a requirement for variables that can store arrays of values. These are sometimes referred to as multi-element variables. An array can consist of multiple elementary data types or other derived data types.

For example, consider a variable required to hold a set of operating pressures for various points around a steam vessel. Suitable data types would be:

```
TYPE VESSEL_PRESS_DATA :
  ARRAY[1..20] OF PRESSURE;
END_TYPE

TYPE VESSEL_MATRIX :
     ARRAY[1..3,1..4] OF VESSEL_PRESS_DATA;
END_TYPE
```

Note: PRESSURE is also a derived type. VESSEL_MATRIX is a two dimensional array of 3 by 4 elements where each element is itself a VESSEL_PRESS_DATA array.

The number of array dimensions and depth of array nesting that is supported, i.e. number of levels at which arrays can exist within other arrays, is not defined in the standard and will vary between different products.

In Structured Text, the value of an array subscript may be given as an integer literal or any expression that produces an integer result.

3.10 Default initial values

All derived data types can be assigned new initial values that will override the defaults of the elementary data types. The new default value is defined with the type definition.

Examples:

```
TYPE  PRESSURE: REAL := 1.0; (* Default 1 bar *)
END_TYPE

TYPE PRESSURE_SENSOR:
  STRUCT
    INPUT: PRESSURE := 2.0; (* Override default *)
    STATUS: BOOL := 0;      (* Default 0 *)
    CALIBRATION:
         DATE := DT#1994-01-10; (* Install Date*)
    HIGH_LIMIT: REAL := 30.0;(* Default limit*);
    ALARM_COUNT: INT := 0;    (* No alarms *)
  END_STRUCT;
END_TYPE

TYPE  DEVICE_MODE:
       (INITIALISING,RUNNING,STANDBY,FAULTY )
   := STANDBY; (* Default enumerated value *)
END_TYPE
```

In array initialisation, each element in the array can be assigned a value. In the following example the first ten elements are assigned the value 1.1, the next five the value 1.3 and the last five the value 1.7.

```
VAR
  VESSEL_PRESS_DATA: ARRAY[1..20] OF PRESSURE
                 := [10(1.1), 5(1.3), 5(1.7)];
END_VAR
```

It is possible to define new defaults for structured data types that are used in new derived data types.

Example:

```
TYPE GAS_PRESSURE_SENSOR :
       PRESSURE_SENSOR( PRESSURE    := 4.0,
                        HIGH_LIMIT := 40.0);
END_TYPE
```

GAS_PRESSURE_SENSOR is a data type derived from PRESSURE_SENSOR with different default initialisation values for PRESSURE and HIGH_LIMIT.

3.11 Variables

Variables can be declared at the beginning of each definition for the different program organisation units (POUs), i.e. programs, function blocks and functions. There are different categories of variables depending on whether the variable is used as an input parameter, as an output parameter, or internally within the POU. Note that input, output and input/output variables provide the external interfaces of POUs such as function blocks.

Global variables are a category of variable that can be declared in programs, resources and configurations. By declaring variables within a POU to be 'external', it is possible to access the value of global variables declared outside the POU. External variables declared in function blocks can reference any global variables defined within a configuration, resource or program that contains the function block.

A list of one or more variables in each category is declared using a particular keyword; the end of each list is terminated with the keyword END_VAR. Within each list, each variable identifier is followed by its data type which may be elementary or derived.

If there are a number of variables of the same data type, a list of identifiers can be given as a list separated by commas.

Example:

```
A,B,C   : REAL;
IN1,IN2 : INT;
```

i.e. three variables of type REAL, and two of type INT.

Internal variables

A list of variables used within a POU (i.e. local variables) is declared using the keyword VAR.

Example:

```
VAR
    AVE_SPEED       : REAL;
    Inhibit         : BOOL;
END_VAR
```

i.e. one internal variable of type REAL and one of type BOOL.

Input variables

A list of variables, which act as input parameters to a POU and are supplied values from external sources, is declared using keyword VAR_INPUT. Input variables will be required for programs, function blocks and functions.

Example:

```
VAR_INPUT
  SetPoint  : REAL;
  Max_Count: USINT;
END_VAR
```

i.e. two input variables (otherwise known as input parameters) of type floating point (REAL) and unsigned short integer (USINT).

Output variables

A list of variables, which act as output parameters to a POU and provide values to be written to external variables, is declared using keyword VAR_OUTPUT. Output variables will be required for programs and function blocks but not functions.

Example:

```
VAR_OUTPUT
          Message    : STRING(10);
           Status    : BOOL;
END_VAR
```

This defines two output variables (otherwise known as output parameters), one a string of 10 characters and the other boolean.

Input/output variables

A list of variables, which act both as input and output parameters and can be modified within the POU, is declared using keyword VAR_IN_OUT. These variables receive values from external variables but can be modified within the POU. Externally the value of these variables is accessible in the same way as other output parameters. However, it is important to note that the value of an input/output variable is stored in a variable that is external to the POU.

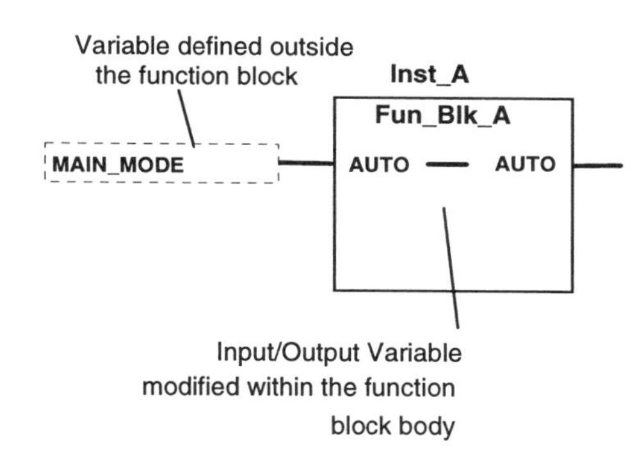

A typical use of this category of variable is for controlling the mode of a function block. Consider a function block with an input/output variable called AUTO. Assume that the function block AUTO parameter is connected to an external variable MAIN_MODE. The value of the initial mode, e.g. INIT, can be written to the MAIN_MODE variable by software outside the function block.

When the function block executes, it has access to the AUTO variable which, in this case, could cause the function block to initialise. The function block can signal the change of mode by writing, say, READY directly to the AUTO parameter. The new mode is accessible either at the function block output AUTO or from the source variable MAIN_MODE.

Input/output variables behave in a manner that is similar to the 'call by reference' feature found in languages such as PASCAL.

When large multi-element variables are passed into a program organisation unit such as a function block, with some implementations it may be more efficient to declare parameters as VAR_IN_OUT rather than using VAR_INPUT or VAR_OUTPUT. VAR_IN_OUT only requires that the address of the variable is passed, rather than copying of the whole variable.

Note: In Function Block Diagram and Ladder Diagram graphical languages, a VAR_IN_OUT variable is depicted as being an input and an output connected by a horizontal line.

Example:

```
TYPE  MODE_LIST :
     ( INIT, READY, RUNNING, STOPPED );
END_TYPE
VAR_IN_OUT
   AUTO : MODE_LIST;
END_VAR
```

Global variables

Global variables can be declared at configuration, resource or program level. They can be accessed by any POU that exists within the configuration, resource or program in which the global is declared using external variables.

For example, consider a configuration LINE_1 that contains a program called PRODUCT_1. Inside the program is a function block called Ramp_Speed. A global, such as Line_Speed declared at the configuration level, can be accessed within function block Ramp_Speed by declaring an external variable that references Line_Speed.

Generally, globals and externals are used to provide access to key process values within programs and function blocks.

Example:

```
VAR_GLOBAL
        Line_Speed      : LREAL;
        Job_Number      : INT;
END_VAR
```

This defines two global variables, one a double precision floating point (LREAL) and one a 16 bit integer.

External variables

External variables can be declared within POUs and provide access to global variables defined at the configuration, resource and program level.

Example:

```
VAR_EXTERNAL
        Line_Speed      : LREAL;
        Job_Number      : INT;
END_VAR
```

i.e. declaration of external variables for referencing two global variables of type double precision floating point (LREAL) and integer.

Temporary variables

The IEC 1131-3 amendment allows temporary variables to be declared within POUs using the VAR_TEMP construct. Such variables will be placed in a temporary memory area, e.g. such as on a stack, which is cleared when the POU invocation terminates.

Example:

```
VAR_TEMP
  RESULT : REAL;
END_VAR;

RESULT := AF18 * XV23 * XV767 + 54.2;
OUT1 := SQRT(RESULT);
```

Directly represented variables

This category of variable allows PLC memory locations[6] to be referenced directly, i.e. without using an identifier. All identities of directly represented variables start with a % character. This is followed by a one or two letter code that defines whether the memory location is associated with inputs, outputs or internal memory, and the type of memory organisation, e.g. as bits, bytes or words.

PLC memory is considered to be organised into three major regions: (I) input locations for receiving values from channels such as analogue and digital input modules, (Q) output locations for values to be sent on to output channels, and (M) internal memory locations for holding intermediate values.

Table 3.7 Directly represented variable codes

First letter code	Interpretation
I	Input memory location
Q	Output memory location
M	Internal memory

Second letter code	Interpretation
X	Bit
B	Byte (8 bits)
W	Word (16 bits)
D	Double word (32 bits)
L	Long word (64 bits)

Note: If the second letter code is not given, the memory organisation is assumed to be as bits.

---

[6] *An amendment to the IEC 1131-3 1993 edition allows directly represented variables to be used within program and function block type definitions. However, such type definitions may not be re-usable on other systems.*

The last part of the identity consists of one or more numeric fields separated by periods (.) to represent the memory location. The way this is interpreted will vary from product to product and is not defined in the standard. However, the numeric fields will be regarded as forming a hierarchical address which can be used to identify hardware sub-divisions, modules, channels etc.

Examples:

```
%I100       (* Input memory bit 100 *)
%IX100      (* Ditto *)
%IW122      (* Input memory word 122 *)
%IW10.1.21  (* This could represent - *)
            (* rack 10, module 1, channel 21 *)
%QL100      (* Output memory long word 100 *)
%MW132      (* Memory location word 132 *)
```

Variable attributes

Variables can also have the following additional attributes: RETAIN, CONSTANT and AT.

RETAIN

This indicates that the values of a list of variables will be retained from when the PLC loses power, to when the power is returned and the PLC warm starts.

Note: With some implementations of PLCs based on IEC 1131-3 , the default is to retain the value of all variables during power failure.

Example:

```
(* Retain the speed profile array  *)
(* and maximum speed on warm starts *)
VAR_OUT RETAIN
  Speed_Profile: ARRAY[1..4] OF REAL;
  Max_Speed REAL;
END_VAR
```

CONSTANT

This attribute indicates that the values of a list of variables cannot be modified. For example, the variable may be initialised with the value of a critical tuning parameter and therefore should not be changed while the program is running. The programming station or compiler should check that the value of such variables cannot be changed. This attribute cannot be used with external variables.[7]

---

[7] *It may be very difficult for a compiler to check for illegal write accesses to a POU in which a CONSTANT is made available via a VAR_EXTERNAL.*

Example:

```
(* Define Startup speed and gear ratio as
constants*)
VAR CONSTANT
  StartUp_Speed: REAL := 12.3; (* Metres/sec *)
  Gear_Ratio:    SINT := 12;
END_VAR
```

Note: The use of VAR_EXTERNAL CONSTANT indicates that the constant value has been declared as a global value externally from the current POU.

AT

This is used to fix the PLC memory location for any particular variable. For example, a variable INPUT_1 may be fixed at input memory location 100 where the value of channel 1 is stored by the PLC hardware input multiplexor. The AT attribute can only be used with global variables and variables declared in programs. Note that variables that do not have the AT attribute, are automatically assigned a memory location by the programming station or compiler.

Example:

```
(* Define scan data array to start at input memory *)
(* location word 10, and digital outputs at        *)
(* output memory bit 120                           *)
 VAR
   SCAN_DATA AT %IW10:    ARRAY[1..8] OF SINT;
   DIG_OUTPUTS AT %QX120: ARRAY[0..15] OF BOOL;
 END_VAR
```

It is also possible to use the AT attribute to declare that particular data types exist in certain memory locations.

Example:

```
(* Declare that an integer exists at input memory
100 and a real exists at output memory 210 *)
 VAR
   AT %IW100: INT;
   AT %QD210: REAL;
 END_VAR
```

Access variables

A list of variables can be declared that provide references by which remote devices and other remote IEC 1131-3 based programs can access certain variables. Access variables provide what the IEC standard calls 'access paths' to named variables. Access variables can only reference the following types of variable:

- Input or output variables of a program;
- Global variables;
- Directly represented variables.

Access variables can reference multi-element data type variables, such as arrays, or a selected element of a multi-element data type variable.

Access variables can have attributes READ_ONLY to indicate that the remote device can only read the referenced variable, or READ_WRITE to indicate that the variable can be both read and written.

Example:

The following access variables provide remote access paths to three variables, LINE1.START_UP, an input parameter of a program called LINE1, SPEED and LENGTH which are global variables. Communications facilities, which are outside the scope of the IEC 1131-3 standard, are able to use the access paths LINE_START_UP, LINE_SPEED and GOOD_CABLE to read and write to LINE.START_UP and SPEED, and read LENGTH.

```
VAR_ACCESS
  LINE_START_UP: LINE1.START_UP: BOOL READ_WRITE;
  LINE_SPEED:    SPEED:          REAL READ_WRITE;
  GOOD_CABLE:    LENGTH:         INT  READ_ONLY;
END_VAR
```

3.12 Variable initialisation

An initial value for a variable can be given in the variable declaration. This will override the default initial value defined for the data type. Since external variables reference global variables they cannot be given initial values.

Examples:

```
VAR
  Process_Runs : INT  := 10;   (*Initial runs = 10 *)
  Max_Temp : REAL := 350.0; (*Initial value 350.0 *)
END_VAR

VAR
  (* Initialise word at input memory *)
  (* location 100 to binary 1001     *)
  AT %I100: WORD := 2#0000_1001;
END_VAR

VAR_OUTPUT
  (* Initial operator message *)
  Message: STRING(12) := 'Operational';
  (* Set-up Speed profile *)
  Speeds: ARRAY[1..4] OF REAL :=
                 10.1, 2(20.5), 33.1;
END_VAR
```

In the following example we use the data type PRESSURE_SENSOR discussed in Section 3.9 'Derived data types'. The example shows how selected elements within a structured variable can be given initial values. These can override default initial values defined in the data type declaration.

```
VAR_GLOBAL
  SENSOR1: PRESSURE_SENSOR(PRESSURE := 4.0,
                     HIGH_LIMIT := 50.0 );
END_VAR
```

3.13 FUNCTIONS

We will now consider the common elements provided to define new types of program organisation units; we will start with functions. A function is a re-usable software element that, when executed with a particular set of input values, always produces a primary value as a result.

Examples are trigonometric functions such as SIN (sine) and COS (cosine), mathematical functions such as SQRT (square root) and text string handling functions such as INSERT (inserts a text string into another text string). Functions can be used in any of the programming languages, for example:

```
Max_Diff_Press := MAX( ABS(Press1A-Press1B),
                       ABS(Press2A-Press2B));
```

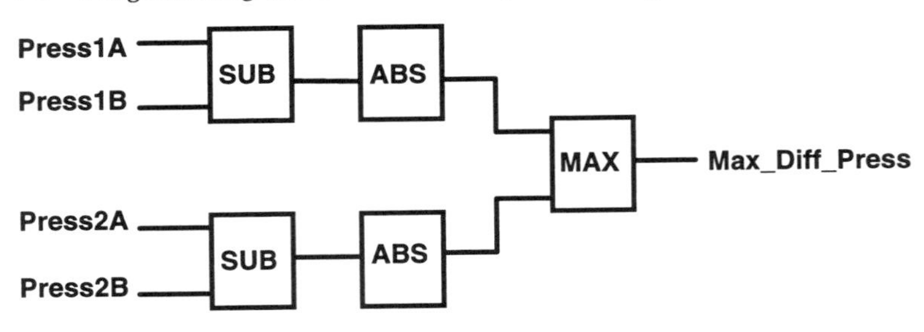

Figure 3.2 Example using functions in ST and FBD

In this example, the maximum differential pressure Max_Diff_Press is calculated from the maximum value of two pressure differences. The same functions are used in the Structured Text language and in the Function Block Diagram. The SUB function is equivalent to subtraction. The ABS function calculates the absolute value, i.e. any negative value is converted into a positive value. The MAX function returns the maximum value of the two inputs.

The important characteristic of all functions is that they cannot store values within internal variables; this contrasts with function blocks that can store values in internal and output variables.

There is a wide range of standard functions which are described in the following sections. Additional user defined functions can be declared using any of the IEC languages. Standard and user defined functions can be used within declarations of other functions, function blocks and programs.

In the following function definitions, where a function has multiple inputs of different data types, a diagram of a function may be shown to clarify how it is used in the graphical languages.

3.14 Function type declarations

New function types can be declared by framing the declaration with the text FUNCTION and END_FUNCTION.

Consider the following example:

```
FUNCTION AVE_REAL : REAL
  VAR_INPUT
    INPUT1, INPUT2 : REAL;
  END_VAR
  AVE_REAL := (INPUT1 + INPUT2)/2;
END_FUNCTION
```

The data type of the result, in this case REAL, is given following the function name. The assignment statement to AVE_REAL creates the value returned by the function.

A function definition contains a list of input and internal variable declarations followed by software that describes the algorithm. This software can be described using one of the following IEC languages :

Structured Text, Function Block Diagram,
Ladder Diagram, Instruction List

The function examples given in this chapter use Structured Text.

An amendment for the second edition of IEC 1131-3 allows inputs to functions to be declared as type VAR_IN_OUT. This implies that variables that are passed as VAR_IN_OUT may be modified within the function body.

Example:

```
FUNCTION INCREMENT : BOOL
 VAR_IN_OUT
   COUNT : INT;
 END_VAR
 VAR_INPUT
   MAX : INT;
 END_VAR

 COUNT := COUNT + 1;
 INCREMENT := (COUNT >= MAX);
END_FUNCTION

(* Example using INCREMENT() function *)
JOBCOUNT : INT;
MaxFlag : BOOL := FALSE;

IF INCREMENT(COUNT := JOBCOUNT,MAX := 100) THEN
     MaxFlag := TRUE;
END_IF;
```

The value of JobCount is incremented within the INCREMENT function body until the value reaches 100.

3.15 Functions for overloaded data types

In Section 3.7 we discussed the hierarchy of elementary IEC data types and associated generic data types such as ANY_NUM. The PLC manufacturer may provide functions that can be used with a variety of different and yet related data types. This avoids the need to have a large number of similar functions each one for a specific data type.

For example, the standard function SQRT (square root) can be used with input values of type ANY_REAL. From Figure 3.1, ANY_REAL is shown as a generic data type which is applicable to both variables of type REAL and LREAL.

The function SQRT is said to support overloaded data type variables. The following Structured Text statements demonstrate how the SQRT function can be used with the two data types.

```
VAR
  Small_Num, Small_sqrt  : REAL;
  Large_Num, Large_sqrt  : LREAL;
END_VAR;

Small_sqrt  := SQRT(Small_Num);
Large_sqrt  := SQRT(Large_Num);
```

Generally the data type of the output of overloaded functions is the same as the input(s). In the previous example, notice that the result of the function is being assigned to the same type as the input parameter.

The data type conversion functions that are discussed in Section 3.17 are an exception to this general rule.

The IEC 1131-3 standard states that the provision of overloaded functions is an optional feature. In practice designing a compiler to handle overloaded variables can be more complex. For PLC products that do not support this option, standard functions for specific data types should adopt the naming convention of appending an underscore '_' and the data type name to the function name.

For example, two forms of the SQRT function, SQRT_REAL and SQRT_LREAL, should be provided for REAL and LREAL data types respectively.

3.16 Invoking functions in Structured Text

When a function is called (i.e. invoked) the function parameter values can be assigned literals, values or expressions. When a function is called, the parameter names come from the identifiers given to the input variables that are declared in the function type definition.

The following Structured Text statements demonstrate how the AVE_REAL function could be called.

```
VAR
  IN1 : REAL := 10.0; (* Initialise variables *)
  IN2 : REAL := 20.0;
  IN3 : REAL := 4.0;
  AVE1, AVE2, AVE3 : REAL := 0.0;
END_VAR

AVE1 := AVE_REAL( INPUT1 := IN1,
          INPUT2 := IN2);(* AVE1 assigned to 15.0*)
AVE2 := AVE_REAL(
          INPUT2 := IN3 + 4.0,
       INPUT1  :=  6.0);(*  AVE2  assigned  to  7.0  *)
AVE3 := AVE_REAL(
     INPUT1 := 4.0 );(* AVE3 assigned to 2.0*)
```

Note 1: In the AVE2 assignment, the order of parameters is reversed. The standard allows parameter names to be given in any order.

Note 2: In the AVE3 assignment, one of the parameters is not used; in this case the unused input parameter will take the default value, i.e. 0.0. Remember that all variables of REAL data type will default to 0.0 unless initialised with a particular value.

Many of the IEC 1131-3 standard functions do not have formal parameter names, in which case the function is called with a list of values. This reduces the need for unnecessary text where parameter names would be superfluous.

Example:

```
VAR
   ALARM_MESSAGE :  STRING(10);
END_VAR
VAR_EXTERNAL
   REPORT:  STRING(15);
END_VAR

REPORT := CONCAT('ALARM',ALARM_MESSAGE);
```

Note 3: CONCAT is one of the standard string functions that is used to concatenate, i.e. join text strings together. It does not have parameter names.

3.17 Data type conversion functions

The IEC 1131-3 languages require strict data type checking. For example, it is not possible to assign the value of a type REAL variable directly to an input of a function of type INT. The programming station or language compiler must always ensure that data types are used consistently. For situations where it is necessary to perform calculations, comparisons and other operations with different data types, the standard defines a wide range of data type conversion functions. The naming convention for such functions is :

```
<Input data type>_TO_<Output data type>
```

Examples:

```
VAR
  COUNT : INT;
  RAMP_RATE : REAL;
  STATUS : WORD;
  DISPLAY_VAL : STRING(16);
END_VAR

(* Convert integer to real *)
 RAMP_RATE := INT_TO_REAL( COUNT );

(* Convert integer to WORD *)
STATUS := INT_TO_WORD( COUNT );

(* Convert a floating point REAL to a
 * string using the normal numeric representation
 *)
 DISPLAY_VAL := REAL_TO_STRING( RAMP_RATE );
```

Floating point (REAL) numbers are converted into integers by rounding the decimal (fractional) portion to the nearest integer. Examples are: 2.6 is rounded to 3, -1.5 is rounded to -2; this is in accordance with IEC 559.

When converting to a string of characters (i.e. STRING) data type, the value will be converted into its normal representation, e.g. integer 12 will be converted to string value '12', a floating point value 154.001 will be converted to string value '154.001'.

Converting binary coded decimal (BCD)

A special set of data conversion functions is also defined for converting binary coded decimal (BCD) values held in bit string data types BYTE, WORD, DWORD and LWORD to integer data types SINT, INT and DINT and vice versa.

Example:

```
VAR
  VALUE : INT;
  BCD_VALUE :WORD := 2#0001_0100_0010; (* BCD 142 *)
END_VAR

(* Convert BCD value to integer *)
VALUE := BCD_TO_INT( BCD_VALUE ); (* value = 142 *)
```

Truncating floating point (REAL) values

A standard function called TRUNC is provided to convert any floating point number to any integer. This function can convert a value of generic data type ANY_REAL to return a value of generic data type ANY_NUM: see Section 3.7 for details on generic data types.

The TRUNC function removes the decimal (fractional) part of the floating point value to produce an integer equivalent to the whole number portion of the decimal.

Example:

```
VAR
  Shaft_rpm      : LREAL;
  Pulse_Count    : INT;
END_VAR

Shaft_rpm := 3200.45;
(* Truncate to produce Pulse Count of 3200 *)
Pulse_Count := TRUNC(Shaft_rpm);
```

Error detection

There can be situations while running a program where data type conversion may not yield sensible values.

For example, when converting a REAL to a SINT, a value that is too large to be stored in a SINT (which is only 8 bits) may be produced. In these situations, the PLC should provide run-time error detection and reporting facilities. However, note that such facilities are not currently defined in the standard.

3.18 Numerical functions

The standard defines a set of numerical functions that includes most of those in common use; they are listed in Table 3.8. All of these functions support data type overloading (see Section 3.7). The result is always the same data type as the input value.

Table 3.8 Numerical functions

Function name	Data type	Description
ABS	ANY_NUM	Absolute value (negative values become positive)
SQRT	ANY_REAL	Square root
LN	ANY_REAL	Natural logarithm
LOG	ANY_REAL	Logarithm
EXP	ANY_REAL	Natural exponential
SIN	ANY_REAL	Sine of input as radians
COS	ANY_REAL	Cosine of input as radians
TAN	ANY_REAL	Tangent of input as radians
ASIN	ANY_REAL	Principal arc-sine, result in radians
ACOS	ANY_REAL	Principal arc-cosine, result in radians
ATAN	ANY_REAL	Principal arc-tangent, result in radians

3.19 Functions equivalent to ST operators

All of the Structured Text arithmetic operators such as '+' and '-' are also provided as standard functions.

These functions fit into two categories, extensible and non-extensible functions. An extensible function is one where the number of inputs can vary depending on the requirements, e.g. the ADD function can add two, three or more inputs. These functions all support overloaded data types as discussed in Section 3.15.

Table 3.9 Extensible arithmetic functions

Function name	Data type	ST operator	Description
ADD	ANY_NUM	+	Addition, result := I1 + I2 + ..
MUL	ANY_NUM	*	Multiplication, result := I1 * I2 * ..

Examples of using extensible arithmetic functions:

```
Fault_Count:= ADD( Dev1, Dev2, AB_34, AB_32, AX_32);
Total_Revs := MUL( Gear1, Gear2, Gear3, 1200 );
```

These function calls are equivalent to the following Structured Text statements:

```
Fault_Count := Dev1 + Dev2 + AB_34 + AB_32 + AX_32;
Total_Revs  := Gear1 * Gear2 * Gear3 * 1200;
```

Table 3.10 Non-extensible arithmetic functions

Function name	Data type	ST operator	Description
SUB	ANY_NUM	–	Subtraction, result := I1 - I2;
DIV	ANY_NUM	/	Division, result := I1 / I2, Note 1
MOD	ANY_INT	**MOD**	Modulus, result := I1 MOD I2, Note 2
EXPT	ANY_REAL Note 3	**	Exponential, result := $I1^{I2}$, Note 3
MOVE	ANY	:=	Result := I1, assigns the value of I1 to the result for any data type

Note 1: With integer division, the result will always be the same data type as the first input parameter I1; any fractional remainder is truncated.

Examples of the integer division (DIV) function are:

```
A := DIV(12,3); (* A is  4 *)
A := DIV(14,5); (* A is  2 *)
A := DIV(-4,3); (* A is -1 *)
```

Note 2: Modulus operation is only valid with integer values and is equivalent to the following Structured Text expression :

```
IF (B1 = 0) THEN
  Result := 0;
ELSE
  Result := I1 - (I1/I2)*I2;
END_IF
```

Examples of the MOD function are :

```
A := MOD(12,3);  (* A is  0 *)
A := MOD(14,3);  (* A is  2 *)
A := MOD(-4,3);  (* A is -1 *)
```

Note 3: The exponential function can only take integer exponents; I2 can only be of type ANY_INT. The result is always the same data type as input 1 (I1).

These functions are provided to allow Structured Text statements involving operators to be expressed in the other languages.

3.20 Bit string functions

These functions provide shift operations for values of bit-string data types. All the functions have the general form :

```
Result := SHL( IN := Input_Bit_String,
                N := Bit_Shifts);
```

The general form of these functions when presented graphically is :

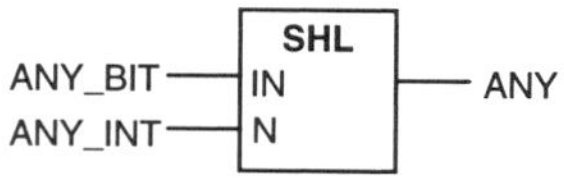

```
            +-----+
            | SHL |
ANY_BIT ----|IN   |---- ANY
ANY_INT ----|N    |
            +-----+
```

Table 3.11 Bit-string functions

Function name	Data type	Description
SHL	ANY_BIT	Shift bit-string n bit positions left, zero fill on the right.
SHR	ANY_BIT	Shift bit-string right n bit positions, zero fill on the left.
ROR	ANY_BIT	Shift bit-string right, rotate by n bit positions.
ROL	ANY_BIT	Shift bit-string left, rotate by n bit positions.

All bit-string shift functions are applicable to overloaded data type ANY_BIT, i.e. they can be used with data types BOOL, BYTE, WORD, DWORD and LWORD.

Example:

```
VAR
  t_8 : BYTE;
  G_8 : BYTE;
END_VAR

t_8  := 2#0011_0101;
G_8  := SHL ( t_8, 4 ); (* G_8 = 2#0101_0000 *)
```

3.21 Boolean bitwise functions

These functions can be used with bit-string values for logical operations. When used with values of boolean (BOOL) data type, they provide the normal logical operations. With data types that have multiple bits, the function applies to bits in matching bit positions in the parameter values. This is best demonstrated by the examples given after the table. These functions all have equivalent Structured Text operators.

With the exception of the NOT function, these are all extensible functions that can take two or more input parameters and support the overloaded data type ANY_BIT.

The standard allows functions to be represented both by function name and optionally by special symbols as shown in the table.

Table 3.12 Boolean bitwise functions

Function name	Data type	Symbol	Description
AND	ANY_BIT	**&**	Result := I1 & I2 &...
OR	ANY_BIT	**>=1** Note 1	Result := I1 OR I2 OR ...
XOR	ANY_BIT	**=2k+1** Note 1	Result := I1 XOR I2 XOR..
NOT	BOOL		Result := NOT I1

Note 1: These symbols can only be used to represent boolean functions in the Ladder Diagram and Function Block Diagram languages; they cannot be used in the textual languages.

Examples of using extensible boolean functions:

```
Confirm := AND( Sw1, Sw2, AB_1, AB_2, AX_2);
Trip_1  := OR ( OverTemp1, OverTemp2, UnderPress );
```

These function calls are equivalent to the following Structured Text statements:

```
Confirm := Sw1 & Sw2 & AB_1 & AB_2 & AX_2;
Trip_1  := OverTemp1 OR OverTemp2 OR UnderPress;
```

Examples of bitwise logical functions:

```
VAR
  Mask   :   WORD := 16#FF01;
  Status:    WORD := 16#0A02;
  X_54   :   WORD;
END_VAR
X_54 := Mask XOR Status; (* X_54 = 16#F503 *)
```

3.22 Selection functions

A wide range of standard selection functions, as listed in Table 3.13, are provided to enable one value to be selected from a number of given values according to certain criteria. These functions can be used with any data type, i.e. generic data type ANY.

The selected value is always the same data type as the inputs; all selection inputs for any particular function call should be of the same data type.

The SEL function is particularly useful as it allows one of two values to be selected depending on a boolean condition.

Examples of selection functions:

```
(* When flag is TRUE, A =230.0 otherwise A =120.0 *)
A:= SEL ( G:= Flag, IN0 := 120.0, IN1 := 230.0 );

(* Select the maximum temperature *)
TempMax := MAX( TempA, TempB, TempC, TempD );

(* Limit soaktime to between 2 and 4 hours *)
SoakTime := LIMIT(MN:=T#2h,IN:=JobTime,MX:=T#4h);

(* Select value from next input channel *)
value := MUX(K:= ChanNo,IN0:=Chan0, IN1:=Chan1,
       IN2:=Chan3);
```

Table 3.13 Selection functions

Function name	Data type	Graphical form	Description
SEL	ANY	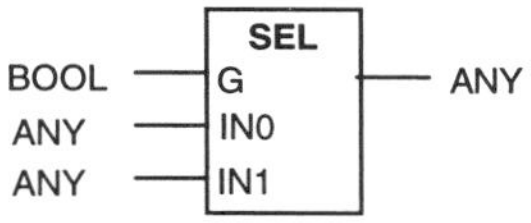 SEL block: BOOL — G, ANY — IN0, ANY — IN1; output — ANY	**Selection** If G = TRUE then Result := IN1 ELSE Result := IN0
MAX	ANY	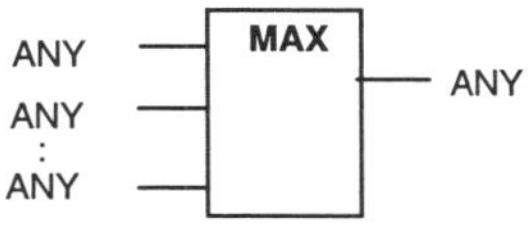 MAX block: ANY, ANY, : , ANY; output — ANY	**Maximum** Result := maximum value of all inputs.
MIN	ANY	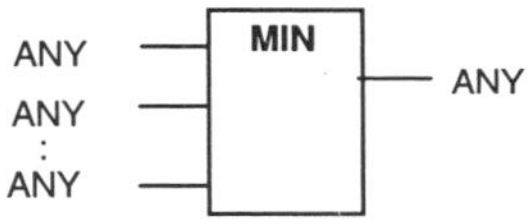 MIN block: ANY, ANY, : , ANY; output — ANY	**Minimum** Result := minimum value of all inputs.
LIMIT	ANY	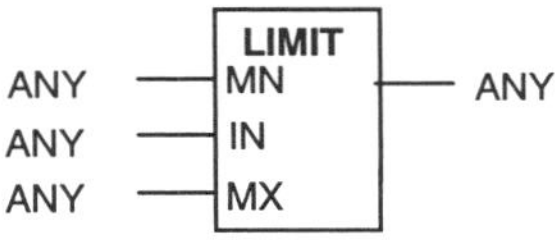 LIMIT block: ANY — MN, ANY — IN, ANY — MX; output — ANY	**Limit** Result is the value of IN limited between a minimum value of MN and a maximum of MX.
MUX	ANY	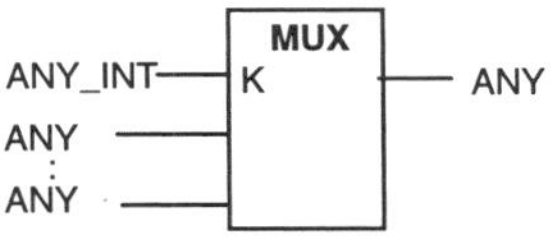 MUX block: ANY_INT — K, ANY, : , ANY; output — ANY	**Multiplexer** Result is the value of the input selected by the value of K (see examples).

3.23 Comparison functions

Comparison functions are provided, as listed in Table 3.14, for comparing values of the same data type. All comparison functions return a boolean (BOOL) data type. These functions can be used with all data types, i.e. generic data type ALL.

Table 3.14 Comparison functions

Function name	Symbol	Description
GT	>	**Greater than** Result := IN1 > IN2
GE	>=	**Greater than or equal** Result := IN1 >= IN2
EQ	=	**Equality** Result := IN1 = IN2
LE	<=	**Less than or equal** Result := IN1 <= IN2
LT	<	**Less than** Result := IN1 < IN2
NE	<>	**Inequality** Result := IN1 <> IN2 Note 1

These functions can also be used as operators in Structured Text, in which case the appropriate symbol should be used.

With the exception of the inequality comparison function, the rest can be used with multiple inputs, in which case the comparison is made between a sequence of inputs and ANDing the result - see the following examples.

These functions can also be used to compare character strings, in which case, when shorter strings are compared with longer strings, the shorter string is considered to be extended with characters which have a code value of zero. The comparison is based on the value of the numeric codes and compares characters from left to right. The standard bases all character codes on the character set defined in ISO/IEC 646.

Examples of comparison functions:

```
(* Set output TRUE if speed 1 less than speed 2 *)
Output_1 := LT ( Speed1, Speed2 );

(* Comparison using Structured Text operator *)
Output_1 := Speed1 > Speed2;
```

Example of using extensible inputs with comparison functions:

```
(* Check that boiler pressures all decrease *)
(* from P1 to P4 *)
PressOK := GT ( P1,P2,P3,P4 );

(* Equivalent Structured Text expression *)
PressOK := P1 > P2  &  P2 > P3  &  P3 > P4;
```

3.24 Character string functions

A comprehensive range of character string functions is defined that enable strings containing textual or non-printable character codes to be manipulated. The use of standard string functions can greatly simplify string handling, for example when constructing operator messages or messages for transmission to remote devices.

Table 3.15 Character string functions

Function name	Graphical form	Description
LEFT	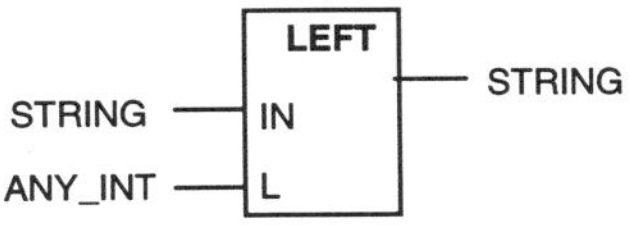 LEFT block: STRING — IN; ANY_INT — L; output — STRING	**Extract left string** Result is the string formed from L characters from the leftmost character of string IN.
RIGHT	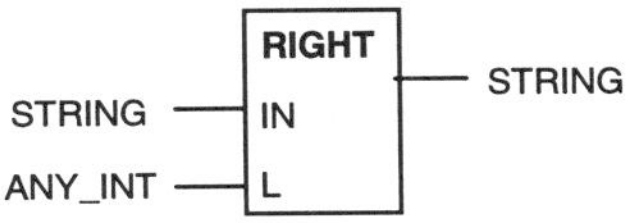 RIGHT block: STRING — IN; ANY_INT — L; output — STRING	**Extract right string** Result is the string formed from L characters from the rightmost part of string IN.
MID	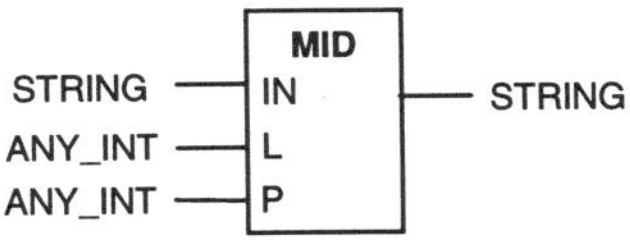 MID block: STRING — IN; ANY_INT — L; ANY_INT — P; output — STRING	**Extract mid string** Result is a string extracted from the input string IN starting at character position P, and L characters long.
CONCAT	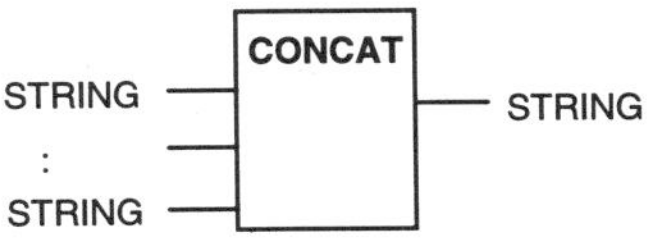 CONCAT block: STRING — ; : ; STRING — ; output — STRING	**Concatenate strings** Result is a string formed by joining the input strings together. This is an extensible function that can take two or more input strings.

Table 3.16 Character string functions - continued

Function name	Graphical form	Description
INSERT	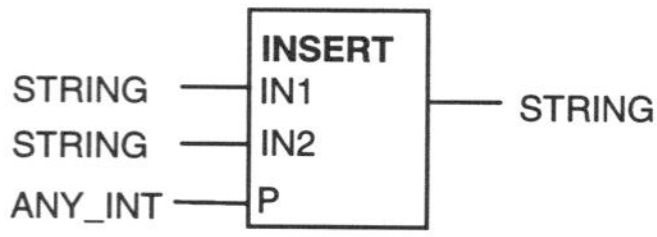 INSERT block: STRING — IN1; STRING — IN2; ANY_INT — P; output — STRING	**Insert string** The result is formed by the string IN2 being inserted into string IN1, P character positions from the start of IN1.
DELETE	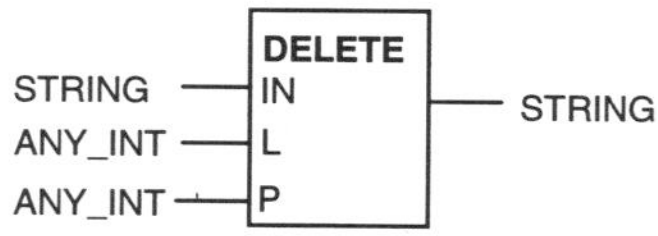 DELETE block: STRING — IN; ANY_INT — L; ANY_INT — P; output — STRING	**Delete string** The result is formed by a string of characters L in length, being deleted from the input string IN, starting from character position P.
REPLACE	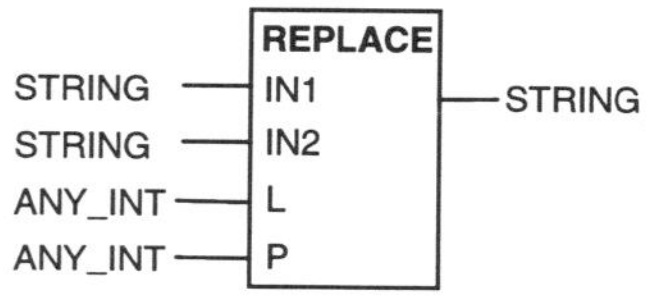 REPLACE block: STRING — IN1; STRING — IN2; ANY_INT — L; ANY_INT — P; output — STRING	**Replace string** The result is formed by replacing L characters in string IN1, starting at position P, with character string in IN2.
LEN	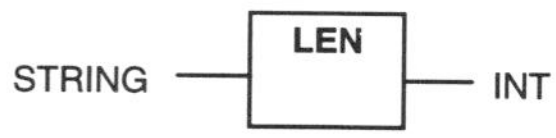 LEN block: STRING —; output — INT	**Length** Result is length of the input string.
FIND	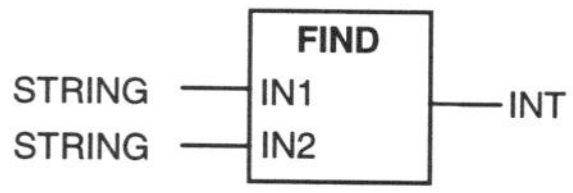 FIND block: STRING — IN1; STRING — IN2; output — INT	**Find string** Result is the position where string IN2 is first found in string IN1. If string IN2 is not found in IN1, the result is 0.

Note 1: It is an IEC convention that the first character in a string is considered to be at position 1. If the length of a string is L, then the last character in the string is at position L.

Note 2: The maximum length of strings that can be supported by a PLC is not defined in the standard.

A range of different errors may occur due to inappropriate values for inputs to these functions. For example, using the CONCAT function could result in producing a string that is beyond the maximum supported string length. The treatment of such errors is not defined in the standard but will clearly need to be addressed in any implementation of IEC 1131-3.

The following example shows how the character string functions can be used to build up typical batch related strings:

```
VAR
  Recipe_Spec : STRING(12) := 'Recipe_x_141';
  Recipe : STRING(14);
  Job_Code : STRING(3);
  Batch : STRING(20);
  BatchId : INT := 7;
END_VAR

(* Extract Job Code 141 *)
Job_Code:= RIGHT ( IN:= Recipe_Spec, L := 3 );

(* Create Recipe name 'Recipe_A7X_141' *)
Recipe := REPLACE( IN1:= Recipe_Spec,IN2:='A7X',
                   L:=1, P:= 8);

(* Build the Batch description*)
(* as 'Recipe_A7X_141_7' *)
Batch := CONCAT(
 Recipe,'_',INT_TO_STRING(BatchId));
```

3.25 Functions for handling times and dates

The following functions can also be used in computations involving durations (TIME), dates (DATE) and times of day (TIME_OF_DAY). These can be particularly useful for time related calculations, such as creating time stamps, determining when sensors need to be calibrated or deciding when items of plant need maintenance. Time handling functions are particularly useful for batch oriented systems.

In the following descriptions, each function is regarded as having two inputs, IN1 and IN2.

ADD (Structured Text + operator)

The ADD function or operator can be used with the following combinations of time and date data types.

IN1	IN2	Result
TIME	TIME	TIME
TIME_OF_DAY	TIME	TIME_OF_DAY
DATE_AND_TIME	TIME	DATE_AND_TIME

For example, adding a duration (TIME) to a time of day (TIME_OF_DAY) will produce an advanced time of day.

SUB (Structured Text – operator)

The SUB (subtract) function or operator can be used with the following combinations of time and date data types.

IN1	IN2	Result
TIME	TIME	TIME
DATE	DATE	TIME
TIME_OF_DAY	TIME	TIME_OF_DAY
TIME_OF_DAY	TIME_OF_DAY	TIME
DATE_AND_TIME	TIME	DATE_AND_TIME
DATE_AND_TIME	DATE_AND_TIME	TIME

For example, subtracting a date (DATE) from another date will produce a duration (TIME) subtracting a duration from time of day will produce an earlier time of day.

*MUL and DIV (Structured Text * and / operators)*

MUL (multiply) and DIV (divide) functions or their equivalent Structured Text operators can be used to scale durations (TIME).

IN1	IN2	Result
TIME	ANY_NUM	TIME

Example:

```
VAR
  TIME : processTime := T#2h;
  TIME : jobTime;
  REAL : scale := 1.5;
END_VAR
```

```
(* Set jobTime to 3.0 hours *)
JobTime := MUL( processTime,scale );

(* Equivalent Structured Text *)
JobTime := processTime * scale;
```

CONCAT

The string concatenation function CONCAT can be used to append a time of day to a date to create a date and time.

IN1	IN2	Result
DATE	TIME_OF_DAY	DATE_AND_TIME

Example:

```
VAR
  DATE : startDate := DATE#1994-03-19;
  TIME : alarmTime := TIME_OF_DAY#13:15:00;
  DATE_AND_TIME : timeStamp;
END_VAR

(* Set time stamp to 1994-03-19-13:15:00 *)
timeStamp := CONCAT ( startDate,alarmTime);
```

3.26 Time related conversion functions

Two special conversion functions are provided to extract times of day and dates from the compound date and time data type.

DATE_AND_TIME_TO_TIME_OF_DAY

This function can be used to extract the time of day from a compound date and time of day.

DATE_AND_TIME_TO_DATE

This can be used to extract the date from a compound date and time of day.

Example:

```
VAR
  DATE_AND_TIME :
      event := DATE#1995-03-20-12:15:00;
  DATE : eventDate;
```

```
    TIME_OF_DAY : eventTime;
  END_VAR
  (* Extract the time of day as 12:15:00 *)
  eventTime :=
   DATE_AND_TIME_TO_TIME_OF_DAY( event );
  (* Extract the date as 1995-03-20 *)
  eventDate := DATE_AND_TIME_TO_DATE( event );
```

3.27 Functions for enumerated data types

Table 3.17 Functions to handle enumerated values

Function name	Structured Text symbol	Description
SEL		The selection (SEL) function can be used to select one of two enumerated values depending on a boolean condition.
MUX		The multiplexor (MUX) function can be used to select one from a number of enumerated values depending on the value of an integer selector.
EQ	=	The equality (EQ) function can be used to test the equality of two enumerated values of the same data type.
NE	<>	The inequality (NE) function can be used to test that two enumerated values are not the same.

Enumerated data types, as discussed in Section 3.9, can only be used with a few of the standard functions or with their equivalent Structured Text operators. This is primarily because an enumerated data type does not have any intrinsic value that can be converted to or compared with a numerical data type.

3.28 Function execution control

If a function is used in either of the graphical languages, Ladder Diagram (LD) or Function Block Diagram (FBD), it is possible to control when the function executes using a special input called EN. When a function is connected in a Ladder Diagram rung or is part of a Function Block Diagram network, the execution enable input EN (a boolean variable) should be set TRUE for the function to be executed. While the EN input is FALSE, the function remains dormant and does not assign a value to its output.

Similarly, each function used in either LD or FBD networks has an extra boolean output ENO which is asserted TRUE when the function execution is successfully completed.

If a function fails to execute owing, for example, to an error such as an arithmetic overflow, or division by zero, the ENO output remains FALSE.

By chaining the ENO output from one function to the EN input of a following function, it is possible to ensure that functions in a chain only succeed in producing a valid result when they all execute without error.

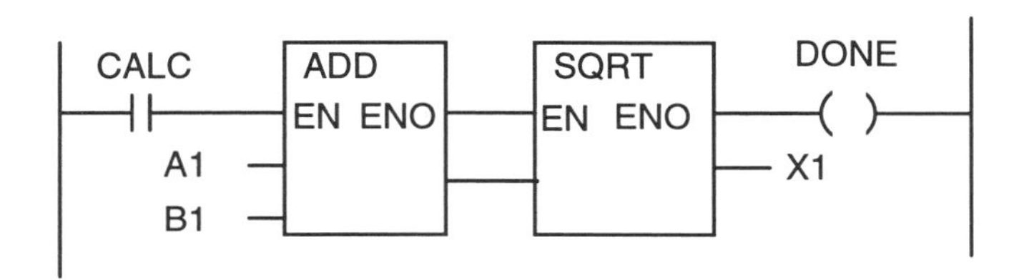

Figure 3.3 EN and ENO example

Figure 3.3 depicts a typical use of the execution control parameters EN and ENO in the Ladder Diagram (LD) language. (See Chapter 6 for further details on LD.)

When the variable CALC becomes TRUE, the ADD function is executed using the current values of A1 and B1. If the ADD function completes successfully, the ENO output is asserted. This in turn enables execution of the SQRT (square root) function which takes the value from the ADD function as input. When the SQRT function completes (assuming the input is a positive number), the ENO output is asserted TRUE. The DONE variable only becomes TRUE when both functions execute successfully and produce the result X1.

All functions may have an EN input and ENO output but in the first revision of IEC 1131-3 these could only be used in the graphical languages LD and FBD. However, an amendment to IEC 1131-3 now also allows their use in Structured Text. Note that the values of EN and EN0 can be changed and tested within function bodies that are written in any of the IEC 1131 languages.

The portion of ladder program depicted in Figure 3.3 could be expressed in Structured Text as follows:

```
IF CALC THEN
  SUM := ADD(A1,B1, ENO => Temp1);
  IF Temp1 THEN
    X1 := SQRT ( SUM, ENO => DONE );
  END_IF;
END_IF;
```

Note1: The => operator is used to transfer the ENO output value to an external variable.

Note2: When expressing graphical links involving functions, additional temporary variables may sometimes be needed when the same functionality is expressed in Structured Text.

When defining a new function type, the EN and ENO variables are implicitly available. In other words, it is not necessary to declare them. However they can be accessed within the function body as if they were normal variables using languages such as Structured Text.

EN and ENO are implicitly defined as follows:

```
VAR_INPUT
  EN : BOOL := 1;
END_VAR
VAR_OUTPUT
  ENO : BOOL;
END_VAR
```

Example setting ENO within a function written in Instruction List:

```
LD   Count        (* Load Count *)
GT   100          (* Is it greater than 100? *)
ST   ENO          (* Set ENO if yes *)
```

Example setting ENO within a function written in Structured Text:

```
ENO := (Count > 100);
```

3.29 FUNCTION BLOCKS

Function blocks are a category of program organisation unit (POU) that allows a specified algorithm or set of actions to be applied to a given set of data to produce a new set of output data. In a control system, function blocks are ideal for algorithms such as PID used for closed loop control, for counters, ramps, filters and so on.

A function block has a defined set of input variables, output variables, and variables for internal storage and temporary data; it also has specific software or an algorithm which behaves as a set of rules, actions or transformations. When a function block executes, it evaluates all its variables; that includes not only the input and internal variables but also the output variables. During its execution, the algorithm creates new values for the output and internal variables.

A function block has a significant feature which is not available with functions, and that is data persistency. Because a function block can hold data values between each execution it can be used to provide a wide range of system building blocks that require retained state.

A function block type is a specification of the data structure and algorithm for a named type of block. A function block instance is a particular set of data values held in structures as defined by the function block type, and that can be transformed by an algorithm that is specified in the function block type definition.

3.30 Function block usage

The following points apply to the use of function blocks:

- It is only possible to access the input and output parameters of a function block instance externally. Internal variables are not accessible.
- A function block instance is only invoked if explicitly requested either because (a) the function block instance is part of a graphical network of connected blocks which form a program organisation unit or (b) it is invoked by a call in the textual languages Structured Text or Instruction List.
- Instances of a particular function block type can be used in other function block or program type definitions.
- Function block instances are always declared using variable declarations within a POU, such as a program or function block. By default, variable declarations are local with the result that function block instances are only visible within the immediate POU. However, function block instances may be declared as globals using the VAR_GLOBAL construct. They are then accessible within any program or function block that exists within the resource or configuration in which the global instances are declared.
- It is possible to pass a function block instance as an input to another program, function block or function. This leads to some interesting and novel possibilities in programming. For example, a tuning function block could in turn be invoked with different instances of different PID function blocks in order to tune a number of control loops.

- The current values of function block input and output parameters can be accessed in the textual languages as if the instance data were held in data structure. For example, the output parameter OUT of function block instance C1 can be accessed as C1.OUT.
- A function block instance can also be represented in the graphical languages Function Block Diagram (FBD) and Ladder Diagram (LD); this is discussed in detail in the following chapters.

3.31 Defining function block types

A function block type definition is introduced by the IEC keyword FUNCTION_BLOCK and terminated by END_FUNCTION_BLOCK. The definition contains a list of input, output and internal variable declarations followed by software that describes the algorithm. This software, sometimes referred to as the function block body, can be described using one of the following IEC languages:

Structured Text, Function Block Diagram, Ladder Diagram, Sequential Function Chart, Instruction List

Function block type definitions may also contain instances of other function blocks and external variables, i.e. that reference global variables declared at the program, resource or configuration level.

The declaration of global and directly represented variables, i.e. those that make reference to absolute PLC addresses, was not permitted in the IEC 1131-3 1993 edition but is allowed in the proposed amendment for the second edition.

IEC 1131-3 defines a number of standard function blocks; these are described in Chapter 9, 'Function blocks'.

3.32 Declaring function block instances

Function block instances can be declared using the same constructs as variables. In effect, a function block instance is really a data structure which has an associated algorithm defined in the function block type definition. Function block instances can only be declared within program or function block type definitions. Creating a copy or instance of a function block from the function block type is called 'instantiation'.

For example assuming that SPEED_RAMP, PRESS_MONITOR are function block types that have already been defined, then examples of instance declarations of these blocks are:

```
VAR
  Line1_Ramp, Line2_ramp : SPEED_RAMP;
END_VAR

(* Instance data retained on warm start *)
VAR RETAIN
  Press_X32, Press_X54 : PRESS_MONITOR;
END_VAR

(* Global function block *)
VAR_GLOBAL
  Main_Press : PRESS_MONITOR;
END_VAR

(* Reference to a function block instance *)
(* Passed into another POU *)
VAR_INPUT
  PressMon : PRESS_MONITOR;
END_VAR
```

3.33 Function block example

Consider the following example of a simple function block called COUNTER. This has one input MODE and one integer output OUT. MODE can only have three values RESET, COUNT and HOLD, i.e. it is an enumerated data type. When the function block is in the RESET mode, the output is always zero; when in the COUNT mode, the output increases by one on each execution; when in the HOLD mode, the value of OUT is held at its last value.

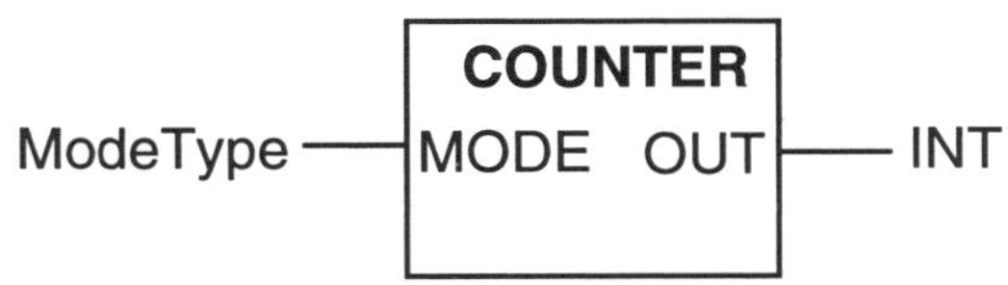

To use the appropriate IEC 1131-3 terminology, when the software associated with a function block type is executed to update the data of a function block instance, the instance is said to have been 'invoked'.

Assume that the definition for the enumerated ModeType is:

```
TYPE
  ModeType : ( RESET, COUNT, HOLD ) := RESET;
END_TYPE
```

The function block type definition for COUNT is:

```
FUNCTION_BLOCK COUNTER
  (* Define the external interface *)
  VAR_INPUT
    MODE : ModeType := RESET;
  END_VAR
  VAR_OUTPUT
    OUT : INT := 0;
  END_VAR

  (* Define the algorithm *)
  IF MODE = RESET THEN
    OUT := 0;
  ELSIF MODE = COUNT THEN
    OUT := OUT + 1;
  END_IF;
END_FUNCTION_BLOCK
```

An instance of the COUNTER function block can be declared within other program organisation units, such as other function blocks. The following fragment of Structured Text shows how an instance C1 of COUNTER could be declared and used.

```
PROGRAM COUNT1
  VAR_INPUT
    InputMode : ModeType;
  END_VAR
  VAR_OUTPUT
    Max_Count : INT;
  END_VAR
  VAR
    C1 : COUNTER;
  END_VAR


 (* Invoke the function block C1 *)
 (* with the mode set according to the value of *)
 (* InputMode, an input parameter to the program*)

  C1(MODE := InputMode);

 (* Assign the output to a variable *)
  Max_Count := C1.OUT;
END_PROGRAM
```

The following table shows an example of states of the input and output parameters of function block C1, which might exist just prior to and after a series of calls to C1, i.e. invocations of C1. The time between invocations will depend on the task configuration and their execution rates which might range from say 10 ms to several hours. The example assumes that function block instance C1 and the parent program execute in the same task.

Notice that the value of InputMode is assigned to MODE just as the function block is invoked. Although the values of output parameters cannot change unless the instance is invoked, the values can always be read and assigned to other variables, in which case the output values are relevant to the last function block invocation.

Table 3.18 Function block invocation states example

Values between invocations time ——>

Variable	Initial state	1	2	3	4	5
InputMode	RESET	RESET	COUNT	COUNT	HOLD	HOLD
C1.MODE	RESET	RESET	RESET	COUNT	COUNT	HOLD
C1.OUT	0	0	0	1	2	2
Max_Count	0	0	0	1	2	2

3.34 PROGRAMS

A program is the largest form of program organisation unit and can be declared at the resource level. A program is very similar to a function block in concept and provides a large re-usable software component. It is defined by a program type definition which, as for function block types, has input, output and internal variable declarations and a body which contains software describing the program's behaviour.

3.35 Program usage

Programs are generally regarded as large software building blocks. A typical PLC project may make use of extensive libraries of function block types, but have relatively few types of program. A program might be used for example to control a major item of plant such as a steam turbine, a reactor vessel or a production cell.

Apart from the following distinguishing features, programs are very similar to function blocks.

- Programs may contain declarations of directly represented variables that allow PLC memory to be accessed using hierarchical memory addresses.
- Programs may contain declarations of global variables. These can be referenced within function blocks using external variables.
- Programs may contain access variables (VAR_ACCESS) that allow named variables to be read and written to by remote devices using communication services.
- They may not contain instances of other programs, i.e. programs cannot be nested.
- Programs may contain instances of function blocks which may optionally be executed under the control of different tasks.
- Instances of programs can only be declared within resources.

3.36 Defining program types

A program type definition is introduced by the IEC keyword PROGRAM and terminated by END_PROGRAM. The definition contains a list of input, output and internal variable declarations followed by software that describes the behaviour of the program. The program body can be described using one of the following IEC languages:

Structured Text, Function Block Diagram, Ladder Diagram, Sequential Function Chart, Instruction List

Program definitions typically contain instances of function blocks, internal variables and external variables, i.e. that reference global variables declared at the resource or configuration level, and directly represented variables.

Example:

```
PROGRAM fermenter
  VAR_INPUT(* Program inputs *)
        Reagent_Code : INT;
        Sterilise : BOOL;
        Ferment_Period : TIME;
  END_VAR
  VAR_OUTPUT (* Program outputs *)
    Yield  : REAL;
        Status : WORD;
```

```
    END_VAR
    VAR (* Internal variables and function blocks *)
      pH_Loop, Temp_Loop : PID;
      Phase : INT := 1;
    END_VAR
    (* Body not shown in example but
     * can be ST, FBD, LD
     * SFC or IL
     *)

END_PROGRAM
```

3.37 Declaring program instances

An instance of a program of a certain type is declared using the keyword PROGRAM followed by the identifier and type of program.

Examples:

```
PROGRAM Line1 : Fermenter;
PROGRAM CT1,CT2 : COUNT1;
```

Line1 is an instance of a program type called Fermenter; CT1 and CT2 are instances of program type COUNT1.

A program declaration in a resource can also contain connections between program inputs and outputs to variables declared outside the program.

Example:

```
PROGRAM Line1 : Fermenter( Reagent_Code := A1,
           Sterilise := A2,
           Ferment_Period := FTIME,
           Yield => AJ_43, Status => KX56 );
```

Program instance Line1 will be run with resource variables or globals A1, A2, FTIME supplying the values for program inputs Reagent_Code, Sterilise and Ferment_Period respectively. Program outputs Yield and Status are written to variables AJ_43 and KX56. These could be declared within the parent resource or as globals at resource or configuration level. Program inputs and outputs can also be connected to directly represented variables, i.e. to PLC memory or hardware references.

In some cases it may be necessary for the program to be assigned to run under the control of a particular task, for example because it must respond to changing plant conditions within a certain deadline. (Tasks are described in Section 3.40.)

The keyword WITH can be used to assign a task to the program instance or to a designated function block instance within the program.

Example:

```
PROGRAM Line2 WITH Slow_task : Packaging_Line,
                Speed := %IW21_10,
                Product_Rate => %QW33_81,
                Sampler1 WITH Fast_Task,
                Sealer1 WITH Fast_Task );
```

Line 2 is an instance of program type Packaging_Line which runs under the control of the Slow_task task. Program input variable Speed is supplied a value that comes from a directly represented variable that might, for example, be associated with a particular PLC input channel. Program output Product_Rate is written to a certain PLC memory reference, that again could be associated with a PLC output channel. Function block instances Sampler1 and Sealer1 declared within the program are defined to run in the Fast_Task.

> Note: An amendment to the IEC 1131-3 1993 edition allows the => operator to be used in both Structured Text program and function block invocations. It indicates that the value of output variable on the left is passed to the variable given on the right of the operator.

3.38 RESOURCES

A resource corresponds to a processing facility that is able to execute IEC programs. A resource is defined within a configuration using the keyword RESOURCE, this is followed by an identifier and the type of processor on which the resource will be loaded. The resource definition that follows contains a list of variable declarations, tasks and program declarations; the definition is terminated with the keyword END_RESOURCE.

> Note: Processor types are not defined in IEC 1131-3 and will vary from product to product; typical types might be PROC_386, LCM_68000.

3.39 Resource usage

A resource may contain the following:

- Global variable declarations;
- Access paths, i.e. access variables that permit remote access to named variables;
- Variable declarations including external variable declarations that reference globals at the configuration level;
- Program declarations;
- Task definitions that can be assigned to programs and function blocks within the resource.

A full example is shown in Section 3.45 'Configurations'.

3.40 TASKS

In the design of a real-time system, the designer requires flexible and configurable control over the execution rates of different parts of the program. Generally the different parts should execute at rates determined by the requisite responsiveness of the system and a need to optimise the use of the PLC processing capacity. For example, in applications handling temperature control of objects of high thermal mass such as in a furnace, the associated function blocks could execute, say, every minute and still provide stable control. In contrast, function blocks involved in interlock logic for, say, a fast acting packaging machine might need to be executed every 5 ms.

Tasks are associated with a particular resource and are considered to execute under the control of the resource.

> Note: IEC 1131-3 assumes that tasks in different resources always execute independently. However some form of timing synchronisation between resources may be provided in some implementations.

The standard allows designated function blocks and programs to be assigned to different tasks. A task is a scheduling facility that executes either periodically or in response to a particular boolean variable changing state. Each task may be assigned an interval, i.e. that defines the time between each execution and a priority (0 is considered highest, then 1, 2 etc.).

3.41 Non-preemptive and preemptive scheduling

Where there are multiple tasks declared, each task is generally assigned a different interval and priority. The moment in time at which a particular task is executed depends on the type of scheduling regime, the priorities and intervals of the other tasks and how long each of the other tasks has been ready and waiting to execute.

A PLC can provide two methods of scheduling - non-preemptive and preemptive. The method adopted can significantly alter the behaviour of the system.

Non-preemptive scheduling

We will consider non-preemptive scheduling first. With this scheduling regime, once a task is executing, it always continues until all the programs and function block instances assigned to it have been executed once. The task then terminates; the next to execute is the waiting task with the highest priority, or if there are several tasks with the same priority, it is the task that has been waiting the longest. Once a task has executed, it is not considered for scheduling again until a period equal to the task interval has elapsed.

Designing an operating system or executive based on non-preemptive scheduling is generally more straightforward than preemptive scheduling. However, non-preemptive systems tend to have poor characteristics for stable control.

The actual real-time interval between task executions may vary considerably as it depends on how long the other tasks take to complete. If one task occasionally takes significantly longer to execute, then all other tasks may be delayed. This can result in erratic task execution, i.e. it is not always possible to predict when certain tasks will execute. Such systems are said to be non-deterministic and generally should not be used for time critical applications.

Preemptive scheduling

Preemptive scheduling is required for deterministic systems, i.e. control systems that are required to have consistent timing relationships between events, for example where certain control algorithms are updated at precise and regular periods. With this regime, irrespective of what task is currently executing, when a higher priority task interval elapses it is immediately scheduled. The currently active task is suspended. When the higher priority task terminates, the suspended task of lower priority continues. Where there are several priority levels, a number of tasks may have started execution but be suspended while higher priority tasks execute.

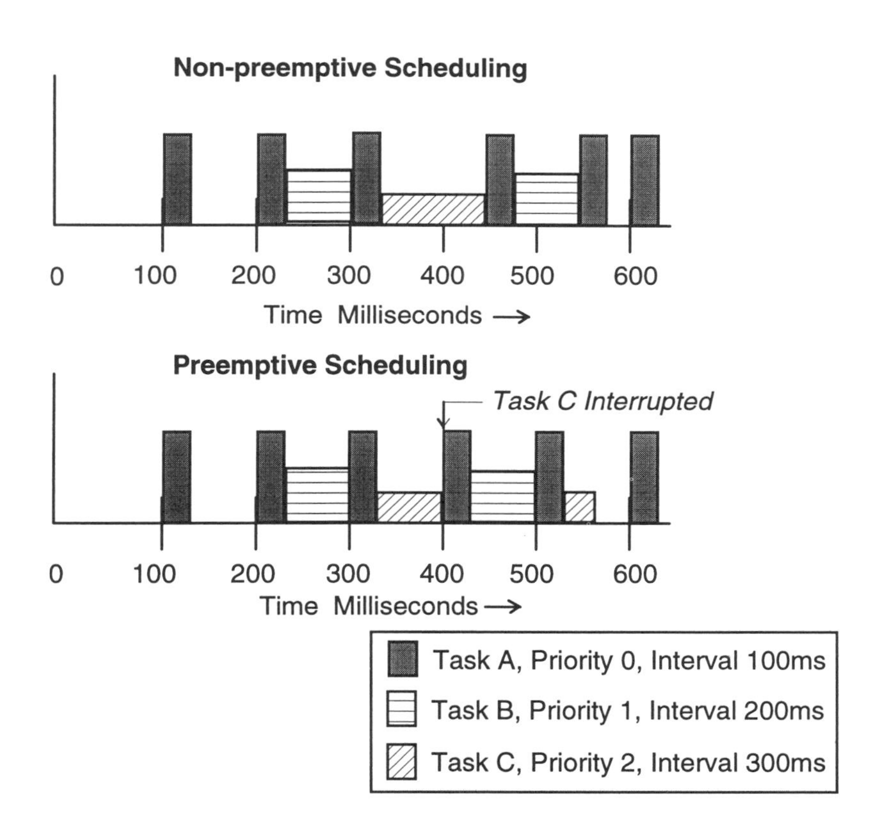

Figure 3.4 Non-preemptive and preemptive scheduling

Figure 3.4 compares the execution timing of three tasks under non-preemptive and pre-emptive scheduling regimes. Tasks A, B and C have priorities 0, 1, 2 respectively (task A with priority 0 is highest). Task A should execute every 100 ms, B every 200 ms and C every 300 ms. With non-preemptive scheduling, notice that tasks A and B are delayed after C executes. Task A is further delayed when it attempts to execute after 500 ms.

In contrast, with pre-emptive scheduling, task A always executes every 100 ms, and B every 200 ms, although it has to wait for the higher priority task to complete. Task C is interrupted to allow first A and then B to execute.

Note: Products that allow resources to support multi-tasking scheduling facilities will require mechanisms to indicate when tasks over-run, i.e. when the time taken to complete a task is longer than the task interval. Ideally programming systems should also offer facilities to measure and optimise processor utilisation.

3.42 Task declarations

A task declaration is introduced using the keyword TASK followed by the task identifier and optional values for the following three parameters:

Table 3.19 Task declaration parameters

Task parameter	Data type	Description
SINGLE	BOOL	A Boolean variable whose value 'on the rising edge', i.e. on changing from 0 to 1, will cause the task to be scheduled for execution once. Note 1
INTERVAL	TIME	The period that should elapse between subsequent task executions. Provided that this parameter is not zero, and the SINGLE parameter is 0, the task will execute periodically. Note 1
PRIORITY	UINT	The task priority level. 0 is regarded as highest, i.e. 0,1,2,3 are decreasing priority levels.

Note 1: The task will not be scheduled while the value of the SINGLE parameter is 1.

Tasks can also be declared graphically; this is discussed in Chapter 5, 'Function Block Diagram' using a special graphical format called 'Continuous Function Chart'.

Note: The number of tasks supported, the range of priority levels and the range of values allowed for task intervals will vary from product to product.

Examples of task declarations are:

```
TASK FAST_INTERLOCKS(INTERVAL :=t#30ms,PRIORITY:=0);
TASK LOG_TASK(SINGLE := LogFlag, PRIORITY:=3);
TASK CONTROL_TASK(INTERVAL := t#500ms,PRIORITY:=1);
```

3.43 Assigning programs and function blocks to tasks

Programs and function block instances can be assigned to different tasks. Arranging parts of a program to execute at different rates may be required to meet certain system performance criteria, like ensuring that particular outputs are always updated within a certain time in response to a given change in input values.

A function block or program is associated with a task using the keyword WITH, as shown in the program declaration example in Section 3.37.

Function block instances that are not assigned to tasks execute in the same task as the parent program.

Any program that is not allocated to a task is by default executed at the lowest priority and will be re-scheduled as soon as it completes, i.e. it runs continually in the background.

3.44 Function block and program execution rules

IEC 1131-3 defines rules for function block and program execution. These are required to ensure that where values are being passed between, say, function block instances that are running in different tasks, the values are always consistent, i.e. are all produced as a result of a particular task execution. The rules are:

- If more than one input value of a function block A comes from outputs of a function block B, then the system should ensure that all the values from B are produced by the same execution of B.

  This is to ensure data concurrency and avoid unpredictable behaviour that could result from A executing with some input values produced by, say, the latest execution of B and some by a previous execution of B.

- In situations where a number of function blocks are assigned to the same task, and all take values from outputs of a function block B running in a different task, then again their input values should all come from the same execution of B.

To summarise, when sets of values are passed from function blocks running in one task, to those in another, the values should all be produced in the same task execution.

The way this is implemented is not specified. However, the rules imply that when a task initially executes, it should first gather and store all the values of function block inputs arising from function blocks in other tasks. The task then executes all its assigned programs and function blocks. New function block output values created during this time need to be held in a temporary data buffer. At the point where the task execution terminates, all new output values are then copied from the temporary buffer into the normal outputs in one operation, i.e. other tasks should not be able to interrupt during the buffer copy.

3.45 CONFIGURATIONS

A configuration defines the software for a complete PLC or programmable control system and will always include at least one but, in some cases, many resources. A configuration is specific to a particular type of PLC product and the arrangement of the PLC hardware. It can only be used to create software for another PLC if the hardware is identical.

Generally, a configuration will correspond to a particular arrangement of PLC hardware which will include specific:

- processing resources (such as processor boards with certain types of micro-processor);
- memory addresses for input and output channels;
- system capabilities (e.g. maximum number of tasks and execution rates).

A configuration is introduced with the keyword CONFIGURATION and is terminated with the keyword END_CONFIGURATION. The definition can either be in a textual format or graphical. The graphical format is described in Chapter 5, 'Function Block Diagram'.

3.46 Configuration example

The following example depicts many of the features of a configuration:

```
CONFIGURATION unit_1_config
  VAR_GLOBAL
    G_speed_setpoint : REAL;
    G_runUp_Time     : TIME;
    G_Log_Event AT %M100 : BOOL;
    G_Log_Data       : ARRAY[1..100] OF INT;
  END_VAR
  RESOURCE Res1 ON Proc_386
    VAR_GLOBAL
      ALARM_FLAG : BOOL;
    END_VAR
    TASK IO_SCAN_TASK
        (INTERVAL:= t#100ms,PRIORITY:= 0);
    TASK CONTROL_TASK
        (INTERVAL:= t#200ms,PRIORITY:= 1);
    TASK PROG_TASK
        (INTERVAL:= t#400ms,PRIORITY:= 2);
    PROGRAM turbine1 WITH PROG_TASK: turbine (
      speed_setpoint := G_speed_setpoint,
      runUp_time     := G_runUp_Time,
      speed_pv       := %ID200,
      actuator_output => %QW310,
      loop1               WITH CONTROL_TASK,
      ramp1               WITH CONTROL_TASK,
      io_scanner1         WITH IO_SCAN_TASK );
  END_RESOURCE

  RESOURCE Res2 ON Proc_8044
    TASK LOG_TASK
        (SINGLE:= G_Log_Event, INTERVAL:= t#0ms);
    PROGRAM logger1 WITH LOG_TASK: log (
                        data => G_LOG_DATA);
    PROGRAM diagnose1 : diagnostics;
  END_RESOURCE

  VAR_ACCESS
    UNIT1_START:      Res1.Turbine1.Start:
                        BOOL READ_WRITE;
    UNIT1_ALARM:      Res1.ALARM_FLAG:
                        BOOL READ_ONLY;
    UNIT1_LOG:  G_Log_Event: BOOL READ_WRITE;
    UNIT1_DATA: G_Log_Data:  ARRAY[1..100] OF
                      INT READ_ONLY;
  END_VAR
END_CONFIGURATION
```

This example assumes that a PLC system has two processors of type PROC_386 and PROC_8044.

There are two resources Res1 and Res2 that exist on the two processors. Res1 contains a program instance turbine1 which has input and output connections to global variables and to direct PLC memory. Notice that within program turbine1, function block instances loop1, ramp1 and io_scanner1 are configured to run under the control of particular periodic tasks.

The global variables e.g. G_speed_setpoint declared at the configuration level, can be accessed within all programs and function blocks.

Res2 contains two programs logger1 and diagnose1. Logger1 runs under the control of an event driver task LOG_TASK. This will only execute when the global variable changes from 0 to 1, i.e. on a rising edge, and then writes out data to a global array G_Log_Data. The program diagnose1 runs continually at low priority in a background task.

A number of access variables are provided so that remote devices can reference certain variables. For example, UNIT1_ALARM is the external reference to ALARM_FLAG, a boolean global variable declared within resource Res1. UNIT1_START provides remote access to an input variable of program Turbine1.

Using communications services, a remote operator station might for example write to UNIT1_START to initiate a turbine run up sequence within the program turbine1 and monitor UNIT1_ALARM to check for alarms. By writing to UNIT1_LOG, a data logging program logger1 is executed once to update log data. This can then be read from the operator station via access variable UNIT1_DATA.

Note: Variables within a resource are identified by a hierarchical name, starting with the resource name followed by the name of the variable or instance within it, e.g. Res1.Turbine1.Start.

Summary

In this chapter we have reviewed most of the common elements of the IEC languages. The main points covered in this chapter can be summarised as follows:

- The standard defines a diverse range of data types and functions suitable for developing most types of application to be found in the industrial control arena.
- It is possible for control software to be built from a range of well defined elements, that include standard data types through to the larger software elements such as functions and function blocks.

- There is support for hierarchical software structures. For example, elementary data types can be used to define user derived data types which in turn can be used in program organisation units such as functions and function blocks.
- Higher 'functionality' functions and function blocks can be readily built up from definitions of more primitive types. This permits either a top-down or bottom-up design approach to software development.

The following chapters will review how the behaviour of program organisation units can be described using different textual and graphical languages.

Chapter 4

Structured Text

This chapter describes Structured Text, a high level textual language, which is the first of the IEC languages in the set:

Structured Text, Function Block Diagram, Ladder Diagram, Instruction List and Sequential Function Chart.

Structured Text or ST can be used to express the behaviour of functions, function blocks and programs. It can also be used within the Sequential Function Charts to express the behaviour of steps, actions and transitions.

In this chapter we will review:

- How to assign values to variables;
- How to create simple and complex expressions;
- The use of arithmetic, boolean and comparison operators;
- Calling functions and function blocks;
- Using language constructs for conditional statements and repeating a number of statements (iteration);
- How to create nested language constructs.

4.1 Structured Text language

Structured Text or ST is a high level language which has a syntax that on first appearance is very similar to PASCAL. Although there are some minor similarities to PASCAL, ST is a distinct language that has been specifically developed for industrial control applications.

The ST language has a comprehensive range of constructs for assigning values to variables, calling functions and function blocks, for creating expressions, for conditional evaluation of selected statements and for iteration, i.e. repeating selected sections of code.

The language statements can be written in a fairly free style where tabs, line feed characters and comments can be inserted anywhere between keywords and identifiers, i.e. wherever a space character is required. For the software developer, ST is fairly straightforward to learn and use. It is a language that is easy to read and understand, particularly when written using meaningful identifiers and well annotated with comments.

Structured Text is a general purpose high-level language for expressing different types of behaviour involving a variety of different types of data. It is particularly useful for complex arithmetic calculations

Note: The IEC 1131-3 standard only defines Structured Text as the language that consists of statements that can be used to assign values to variables, for example within function block and program bodies. However, many implementors have come to regard all the textual language as defined in Chapter 3 which is used to define common elements, such as variable declarations, as also being part of the Structured Text language.

4.2 Assignment statements

Assignment statements are used to change the value stored within a variable or the value returned by a function. An assignment statement has the following general format:

```
X := Y;
```

where Y represents an expression that produces a new value for the variable X when the assignment statement is evaluated. The value assigned should be of the same data type as the variable.

The expression can be very simple, e.g. a literal constant, or complex, e.g. involving many nested expressions. The assigned variable can be a simple or multi-element variable such as an array or structure.

The following fragment of Structured Text contains a variety of assignment statements:

```
TYPE Alarm
  STRUCT
     TimeOn   : DATE_AND_TIME;
     Duration : TIME;
  END_STRUCT;
END_TYPE

VAR
```

```
    Rate,A1    : REAL;
    Count      : INT;
    Alarm      : Alarm;
    Alarm2     : Alarm;
    Profile    : ARRAY[1..10] OF REAL;
    RTC1       : RTC; (* Real time clock *)
  END_VAR

Rate := 13.1;          (* Literal value i.e. constant *)
Count := Count + 1;(* Simple expression *)
A1 := LOG(Rate);       (* Value from a function *)
Alarm1.TimeOn:= RTC1.CDT;(* Value from a function *)
                          (* block output parameter*)
Alarm2 := Alarm1;      (* Multi-element variable *)

(* Value from a complex  expression assigned to *)
(* a single element of an array *)
Profile[3] := 10.3 + SQRT(( Rate + 2.0 ) *
                 ( A1 / 2.3 ));
```

Note: The PLC operating system should always ensure that assignment operations are completed, i.e. are atomic operations. For example, when an assignment involves copying the values for a multi-element variable, such as from one array variable to another, it should not be possible to access any array element until the array has been completely updated.

4.3 Expressions

Expressions are used to calculate or evaluate values derived from other variables and constants. Structured Text expressions always produce a value of a particular data type, which can be either an elementary or a derived data type. An expression can involve one or more operators, variables and functions.

Composite expressions can be built up by nesting simpler expressions. This allows ST to be used to perform complex arithmetic computations and data manipulations involving different types of data.

Note: The programming station or ST compiler should always ensure that expressions will produce values of the correct data type to match the variable being assigned.

4.4 Operators

The standard defines a range of operators for arithmetic and boolean operations as listed in Table 4.1.

Table 4.1 Structured Text operators

Operator	Description	Precedence
(...)	Parenthesised expression	Highest
Function(...)	Parameter list of a function, function evaluation	
**	Exponentiation, i.e. raising to a power	
- NOT	Negation Boolean complement, i.e. value with opposite sign	
* / MOD	Multiplication Division Modulus operation	
+ -	Addition Subtraction	
<, >, <=, >=	Comparison operators	
= <>	Equality Inequality	
AND, &	Boolean AND	
XOR	Boolean exclusive OR	
OR	Boolean OR	Lowest

Note: The standard allows each operator to have an equivalent function. For the arithmetic operators' equivalent functions, see Chapter 3, Section 19; similarly for the boolean operators' equivalent functions, see Chapter 3, Section 21. The behaviour of the equivalent function is identical to that of the related operator.

Examples of ST operator usage:

```
Area := PI * R * R;
V := K ** (-W * T); (* K to the power of -W * T *)
Volts := Amps * Ohms;
Start := OilPress AND Steam AND Pump;
Status:= (Valve1 = Open) XOR (Valve1 = Shut);
```

4.5 Expression evaluation

Expressions are evaluated in a particular order depending on the precedence of the operators and other sub-expressions. Bracketed i.e. parenthesised expressions have the highest precedence, followed by functions. Operators of the highest precedence are evaluated first, followed by lower precedence operators, down to the lowest. The precedence of the different types of operator is shown in Table 4.1, going from the bracketed expression, highest, down to the OR operator, lowest. Where operators have the same precedence, they are then evaluated left to right.

Consider the following example:
(Assume Speed1 = 50.0, Speed2 = 60.0, Press = 30.0)

```
Rate := Speed1/10.0+Speed2/20.0-SQRT(Press+6.0);
```

Evaluation order:

```
Speed1/10.0       = 5.0
Speed2/20.0       = 3.0
Press + 6.0       = 36.0
SQRT(Press+6.0)   = 6.0
5 + 3             = 8.0 (Speed1/10.0 + Speed2/20.0)
8.0 - 6.0         = 2.0 (8 - SQRT(Press+6.0))
Rate is assigned the value 2.0
```

Consider the evaluation order with the addition of brackets added as shown:

```
Rate :=Speed1/10.0
           + Speed2/(20.0 - SQRT(Press+6.0));
```

Now the expression (20 - SQRT(Press+6.0)) has higher precedence and will be evaluated before its value is used as a divisor for Speed2.
The value for Rate in this case will be:

```
Rate = 5 + 60/(20.0 - 6.0)
Rate = 5 + 60/14.0
Rate = 9.287
```

With boolean operators, the expression is only evaluated up to the point where the value can be resolved.

For example consider the following expression (A, B, D and E are boolean variables):

```
StartUp := A AND B AND D AND E;
```

If A is false, then StartUp will be assigned the value of FALSE, and the rest of the expression will not be evaluated.

4.6 Statements

The Structured Text language has a range of constructs for calling function blocks, conditional evaluation of statements and iteration, i.e. repeating sections of code.

4.7 Calling function blocks

A function block instance can be invoked (i.e. the code body can be executed to update the data associated with the instance) by calling the function block instance name with suitable input parameter values. If parameter values are not provided, the current input parameter values as remaining from the last invocation are used.

If the function block instance has not been invoked before, any input parameters that are not assigned will take default values as given in the function block type definition.

The function block instance invocation has the following general form:

```
FunctionBlockInstance(
        InputParameter1:=ValueExpression1,
          InputParameter2:=ValueExpression2, ...);
```

The ValueExpression should create a value of a data type that matches the function block input parameter. The values can be produced by expressions of any complexity.

Example:

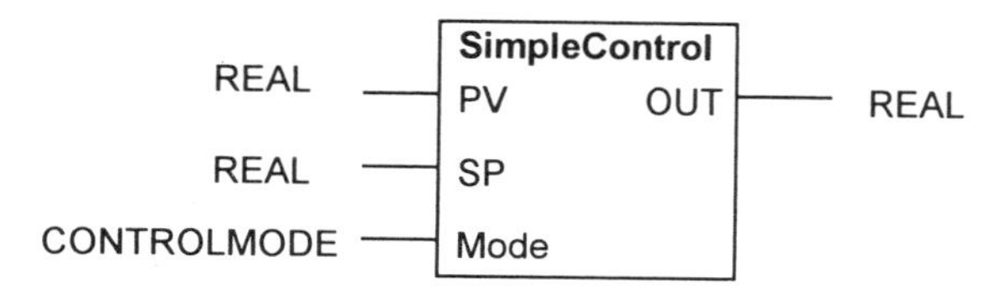

Consider an instance 'Loop1' of a function block type called SimpleControl, which could be, for example, a simple type of PID controller. The Mode input parameter can take two values AUTO, MANUAL, i.e. it is an enumerated type; assume the default value is AUTO.

Fragment of Structured Text calling the function block Loop1:

```
VAR
     Loop1 : SimpleControl;
END_VAR

(* Initial invocation *)
Loop1(PV := Input1.Out + Offset, SP := 100.0);

(* Further code .... *)

(* Subsequent invocation *)
Loop1(PV:= Input1.Out + Offset, Mode:= MANUAL);
```

In the first invocation of Loop1, parameter Mode is not assigned, and therefore its default value AUTO is taken. In the second invocation, the parameter SP is not assigned; therefore the value as remaining from the last invocation is used, i.e. 100.0.

Note: The standard states that assigning values to function block inputs outside a function block invocation is not permitted. However a number of implementations of IEC 1131-3 allow such assignments.
For example: Loop1.Mode := Manual;
The new value assigned to Mode is used the next time Loop1 is invoked. Being able to assign values to function block input parameters ahead of invocation is a useful facility. However such assignments may not be portable to other products.

The values of function block instance output parameters are always available and can be assigned to other variables.

Example:

```
FlowRate := Loop1.Out;
```

This will assign FlowRate with the value generated by the last invocation of the Loop1 instance.

An amendment to the IEC 1131-3 1993 edition allows the value of an output variable to be directly assigned to a variable using the => operator.

Example:

```
VAR
     X12 : REAL;
END_VAR

Loop1(PV:= %IW100, SP := 210.0,OUT => X12);
```

4.8 Conditional statements

Conditional statements are provided so that selected statements will be executed when certain conditions exist.

IF ... THEN ... ELSE

Selected ST statements can be evaluated depending on the value returned by a boolean expression using the IF ... THEN construct. This takes the general form:

```
IF <boolean expression> THEN
  <statements...>
END_IF;
```

The boolean expression can be any expression that returns a TRUE or FALSE boolean result, e.g. the state of a single boolean variable or a complex expression involving numerous variables.

Alternative statements can be executed using the IF ... THEN ... ELSE construct in the form:

```
IF <boolean expression> THEN
  <statements...>
ELSE
  <statements...>
END_IF;
```

Examples of conditional execution:

```
IF Collision THEN
  Speed := 0;
  Brakes := ON;
END_IF;

IF (Gate = CLOSED) AND
     (Pump = ON) AND (Temp > 200.0) THEN
  Control_State := Active;
ELSE
  Control_State := Hold;
  PumpSpeed := 10.0;
END_IF;
```

As IF ... THEN and IF ... THEN ... ELSE constructs are also statements, they can be nested within other conditional statements to create more complex conditional statements.

For example:

```
IF FlowRate > 230.0 THEN
  IF FlameSize > 4.0 THEN
    Fuel := 4000.0;
  ELSE
    Fuel := 2000.0;
  END_IF;
ELSE
  Fuel := 1000.0;
END_IF;
```

Further statements can be conditionally executed within the IF...THEN construct, using the ELSIF ... THEN... ELSE construct, which has the general form:

```
IF <boolean expression> THEN
  <statements...>
ELSIF <boolean expression> THEN
  <statements...>
ELSE
  <statements...>
END_IF;
```

Any number of additional ELSIF sections can be added to the IF ... THEN construct.

For example:

```
IF A>B THEN
  D := 1;
ELSIF A = B +2 THEN
  D := 2;
ELSIF A = B - 3 THEN
  D := 4;
ELSE
  D := 3;
END_IF;
```

CASE

The CASE conditional statement is provided so that selected statements can be executed depending on the value of an expression that returns an integer result e.g. the value of a single integer variable or the integer value from a complex expression. The integer expression should return a value of data type INT.

The set of statements which have an integer selector value that matches the value of the integer expression are executed. If no match is found, the statements preceded by ELSE are executed.

The CASE construct has the following general form:

```
CASE <integer expression> OF
  <integer selector value1> : <statements...>
  <integer selector value2> : <statements...>
   ...
  ELSE
  <statements...>
END_CASE;
```

The <integer selector value> can define one value, several values or a range of values[1].

```
1:    (* Value 1 *)
2,3,4:(* Values 2,3 and 4 *)
5..10:(* Values 5 through 10 *)
```

Example:

```
CASE speed_setting OF
  1 :       speed := 10.0;
  2 :       speed := 20.4;
  3 :       speed := 30.0; fan1 := ON;
  4,5 :     speed := 50.0; fan2 := ON;
  6..10 :   speed := 60.0; water := ON;
ELSE
            speed := 0; SpeedFault := TRUE;
END_CASE;
```

An amendment to the IEC 1131-3 1993 edition allows enumerated variables to be also used in CASE statements.

Example:

```
TYPE
   SPEED: (STOPPED, SLOW, MEDIUM, FAST);
END_TYPE

VAR
  pumpSpeed : SPEED;
  rate : REAL;
```

---

[1] *The .. construct is used to indicate a range of integer values, e.g. 5..10 means 5 to 10.*

```
  alarm AT %Q141 : BOOL;
END_VAR

CASE pumpSpeed OF

  STOPPED : rate := 0.0;
  SLOW    : rate := 20.4;
  MEDIUM  : rate := 30.0;

ELSE

  rate := 0; alarm := TRUE;

END_CASE;
```

4.9 Iteration statements

Iteration statements are provided for situations where it is necessary to repeat one or more statements a number of times depending on the state of a particular variable or condition.

Care should be taken when using iteration statements to avoid situations that result in creating endless loops. Iteration statements may also significantly increase the time to execute different software elements, such as function blocks.

FOR ... DO

The FOR ... DO construct allows a set of statements to be repeated depending on the value of an iteration variable. This is an integer variable of data type SINT, INT or DINT which is used to count the statement executions. This construct takes the general form:

```
FOR <initialise iteration variable>
  TO <final value expression>
  BY <increment expression> DO
  <statements...>
END_FOR;
```

The FOR ... DO construct can be used to count iterations counting either up or down and using any size increment until a final value is reached. The BY keyword is optional; if omitted, the iteration variable will increase by 1.

The test to check whether the value of the iteration variable has reached the final value is always made before executing the statements. It is therefore possible for the final increment to cause the value of the iteration variable to exceed the final value, in which case no further statements within the FOR ... DO will be executed. The standard states that the final value of the iteration variable is implementation dependent. It is therefore recommended that its value is not used outside the FOR loop.

The statements within a FOR ... DO construct should not modify variables that will affect the expressions for the final and increment values.

Examples:

```
FOR I:= 100 TO 1 BY -1 DO
  channel[I].status := ON;
END_FOR;

FOR T := FarmSize-1 TO TankMax*2 DO
  tankNo := T; vessel(tank:= T);
END_FOR;
```

WHILE ... DO

The WHILE ... DO construct allows one or more statements to be executed while a particular boolean expression remains true. The boolean expression is tested prior to executing the statements; if it is false the statements within the WHILE ... DO are not executed. This construct takes the general form:

```
WHILE <boolean expression> DO
 <statements...>
END_WHILE
```

Example:

```
WHILE Value < (MaxValue - 10.0) DO
  bridge1();
  Value := Value + bridge1.Position;
END_WHILE
```

REPEAT ... UNTIL

The REPEAT ... UNTIL construct allows one or more statements to be executed while a particular boolean expression remains true. The boolean expression is tested after executing the statements; if it is true, the statements within the REPEAT ... DO are executed again. This construct takes the general form:

```
REPEAT
 <statements...>
UNTIL <boolean expression>
END_REPEAT
```

Example:

```
tries := 0;
REPEAT
  tries := tries + 1;
  switchgear1(Mode := DISABLE);
UNTIL (switchgear1.State = OFF) OR (tries > 4)
END_REPEAT
```

EXIT

This statement can only be used within iteration statements and allows iteration loops to be ended prematurely. When an EXIT statement is reached, execution continues immediately from the end of the iteration construct; no further part of the iteration construct is executed.

Example:

```
Fault := FALSE;
FOR I:= 1 TO  20 DO
  FOR J := 0 TO 9 DO
    IF FaultList[I,J] THEN
      FaultNo := I*10 + J;
      Fault := TRUE; EXIT;
    END_IF;
  END_FOR;
  IF Fault THEN EXIT; END_IF;
END_FOR;
```

In this example, a two-dimensional boolean array FaultList is scanned. If any element is TRUE, the scan is terminated. An EXIT statement is used in each FOR loop to terminate the loop when a fault is found.

RETURN

This statement can only be used within function and function block bodies and is used to return prematurely from the code body. The RETURN statement causes execution to continue from the end of the function or function block body. [2]

---

[2] *It is recommended that the RETURN construct is used sparingly. The behaviour of function and function blocks with multiple exit paths may be difficult to understand.*

Example:

```
FUNCTION_BLOCK TEST_POWER
  VAR_INPUT
    CURRENT, VOLTS1, VOLTS2, VOLTS3 : REAL;
  END_VAR
  VAR_OUTPUT
    OVERVOLTS : BOOL;
  END_VAR

  IF VOLTS1*CURRENT > 100 THEN
    OVERVOLTS := TRUE; RETURN;
  END_IF;
  IF VOLTS2*(CURRENT+10.0) > 100 THEN
    OVERVOLTS := TRUE; RETURN;
  END_IF;
  IF VOLTS3*(CURRENT+20.0) > 100 THEN
    OVERVOLTS := TRUE;
  END_IF;
END_FUNCTION_BLOCK;
```

In this example, if any test fails, the output OVERVOLTS is set true and the RETURN statement called to terminate the function block execution.

4.10 Implementation dependencies

The IEC 1131 standard sets no limits on the level of complexity that can be expressed in a set of Structured Text statements. However, the standard does state that some implementations may limit certain aspects of the language. For example, there may be limitations on the length of expressions, length of statements, length of comments, number of CASE selections and so on. These limitations may arise because of ST compiler restrictions or because of performance constraints in the target programmable controller.

A list of language features that may have implementation dependent limits is given in an annex at the end of the IEC 1131-3 standard. Because each product may apply different limits to these language features, this may affect the portability of Structured Text between products.

4.11 Structured Text examples

The following examples demonstrate some of the language features that have been reviewed in this chapter.

4.12 Solving a quadratic equation

The following fragment of Structured Text calculates the two roots of a quadratic equation; assume that all the variables are of the REAL data type.

```
tempA := b*b - 4*a*c;
IF tempA > 0.0 THEN
    tempB := SQRT(tempA);
    x_Root1 := (- b + tempB)/2*a;
    x_Root2 := (- b - tempB)/2*a;
ELSE
    msg := 'Imaginary Roots';
END_IF;
```

Note: In a real time control system, floating point calculations such as the square root function shown in this example can use up significant processing time. Where possible, save intermediate results in temporary variables to avoid recalling identical code. It is also advisable to ensure that illegal operations such as taking the square root of a negative number do not occur.

4.13 Calculating average, max and min values

The following ST example calculates the average, and the maximum and minimum values of a set of values read in from a set of input channels.

```
TYPE T_Channel:
   STRUCT
     value : REAL; state : BOOL;
   END_STRUCT;
END_TYPE
TYPE
   T_Inputs: ARRAY[1..32] OF T_Channel;
END_TYPE

VAR
 max: REAL :=0;   (*Set to minimum possible value*)
 min: REAL :=1000.0;(*Set to max. possible value *)
 input : T_Inputs AT %IW130; (* Input channels *)
 sum: LREAL :=0.0;      (* To hold values total *)
 average: LREAL;
END_VAR

FOR I:= 1 TO 32 DO
   sum := REAL_TO_LREAL(input[I].value) + sum;
   IF input[I].value > max THEN
      max := input[I].value;
   END_IF;
   IF input[I].value < min THEN
```

```
        min := input[I].value;
    END_IF;
END_FOR;
average := sum / 32.0;
```

The FOR...DO statement is used to scan the values of all the input channels to calculate the total value for the average, and the maximum and minimum values. The sum is held in the double precision LREAL to improve the accuracy of the result.

Note: This example could also be written using the selection functions MAX() and MIN() as described in Chapter 3, Section 22.

4.14 Tank control function block

The following example describes how ST statements can be used to define the body, i.e. the algorithm of a function block. Consider a function block that controls a tank used to process some type of liquid, which might typically be in part of a food manufacturing or pharmaceutical process. The tank can be filled and emptied using fill and empty valves as depicted in Figure 4.1. The weight of the tank is monitored by a load cell.

The function block monitors the tank Weight to determine whether the tank is full or empty by comparing the Weight with two inputs FullWeight and EmptyWeight.

The block provides a Command input which can have four modes, 1 to fill the tank, 2 to hold the contents, 3 to activate a stirrer and 4 to empty the tank. If the tank is in the correct State the appropriate valves are opened or shut to control the tank level.

The stirrer can only operate if the tank is full; otherwise the command is ignored.

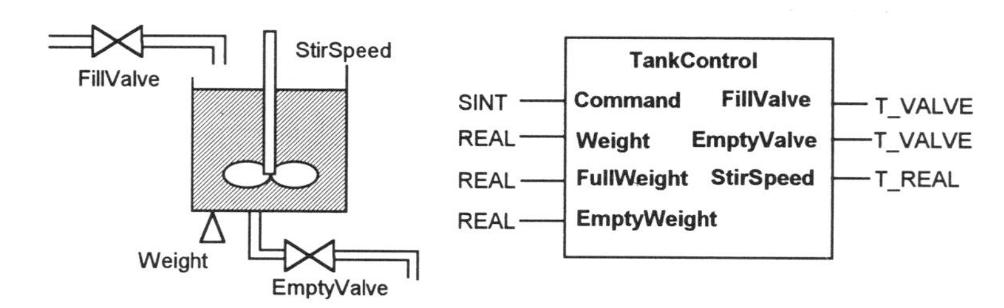

Figure 4.1 Tank control function block example

```
(* Vessel State *)
TYPE T_STATE: (FULL,NOT_FULL,EMPTIED); END_TYPE

(* Valve State *)
TYPE T_VALVE: (OPEN,SHUT); END_TYPE

FUNCTION_BLOCK TankControl

  VAR_IN                      (* Input parameters *)
    Command : SINT;
    Weight : REAL;
    FullWeight, EmptyWeight : REAL;
  END_VAR

  VAR_OUTPUT                  (* Output parameters *)
    FillValve : T_VALVE := SHUT;
    EmptyValve: T_VALVE := SHUT;
    StirSpeed : REAL := 0.0;
  END_VAR

  VAR                         (* Internal variables *)
State : T_STATE := EMPTIED;
  END_VAR

  (* Function Block Body *)

  (* Check the vessel state *)
  IF Weight >= FullWeight THEN  (* Is it full? *)
    State := FULL;
  ELSIF Weight <= EmptyWeight THEN (* or empty?*)
    State := EMPTIED;
  ELSE
    State := NOT_FULL;
  END_IF;

  (* Process the command mode *)
  CASE Command OF
  1 : EmptyValve := SHUT;       (* FILL Tank *)
      FillValve  := SEL(G:= State = FULL,
                     IN0:= OPEN, IN1:= SHUT);
  2 : EmptyValve := SHUT;       (* HOLD contents*)
      FillValve  := SHUT;
  4 : FillValve  := SHUT;       (* EMPTY Tank *)
      EmptyValve := OPEN;
  END_CASE;

  (* Control the stirrer speed *)
  StirSpeed := SEL(G:=((Command = 3)
                 AND(State = FULL));
                 IN0:= 0.0; IN1:=100.0);

END_FUNCTION_BLOCK
```

The first IF... ELSIF statement in the function block body monitors the weight of the tank and sets the tank state, i.e. Full, Not_Full or Empty. The CASE statement checks the command mode and selects the appropriate valve settings.

Notice that the FillValve is set according to the comparison expression 'State = FULL'. While the state is not FULL, this expression is false (i.e. 0) and FillValve is set to the value of select function (SEL) input IN0, i.e. OPEN. As soon as the state is FULL, the expression becomes true (i.e. 1) and FillValve is then set to the value of input IN1, i.e. SHUT.

A further select function ensures that the StirSpeed is only set to 100.0 when the Command mode is 3 and the tank state is FULL.

Typically a function block such as this will be associated with a task so that the algorithm is re-evaluated, say, every 100 milliseconds.

In a real application additional code would generally be needed to check for jammed valves, filter the weight signal which may be noisy due to liquid movement, check for abnormal changes in tank weight and so on.

Summary

In this chapter we have reviewed all the features of the Structured Text language. The main features can be summarised as:

- ST is an expressive language that can be used for programming a wide range of industrial applications.
- It is a high level language with strong data type checking and a formal syntax[3].
- ST can be used to assign values to variables of different data types.
- Values can be calculated using expressions that can be simple or complex.
- The expressions can involve functions or values from function block outputs.
- ST has facilities for conditional evaluation of statements, for repeating sections of code and for calling functions and function blocks.
- ST can be used to describe the behaviour of program organisation units, i.e. functions, function blocks and programs.

---

[3] *A full set of production rules that define the syntax of all textual components of the languages, including Structured Text, is given in Annex B of the IEC 1131-3 standard.*

Chapter 5

Function Block Diagram

This chapter describes Function Block Diagram, the first graphical language of the IEC languages in the set:

Structured Text, Function Block Diagram, Ladder Diagram, Instruction List and Sequential Function Chart.

Function Block Diagram or FBD can be used to express the behaviour of functions, function blocks and programs as a set of interconnected graphical blocks. It can also be used within the Sequential Function Charts to express the behaviour of steps, actions and transitions.

FBD is based on viewing a system in terms of the flow of signals between processing elements. This is very similar to signal flows depicted in electronic circuit diagrams.

In this chapter we will review:

- Common graphical symbols and conventions that apply to FBD and also to the other graphical languages, i.e. Ladder Diagram and Sequential Function Chart;
- How to depict both simple and complex expressions;
- How to show signal flow between functions and function blocks;
- How to assign values to variables;
- How to handle signal feedback;
- The rules for function block evaluation;
- Restrictions on the portability of code between ST and FBD.

5.1 FBD graphical elements

The standard defines types of lines and interconnections for all the graphical languages, i.e. for Function Block Diagram, Ladder Diagram and Sequential Function Chart. IEC 1131-3 allows normal characters such as _ and | to be used to depict graphical objects; this is called semi-graphical representation. However, the standard also allows full graphical representation. The use of full graphics has been adopted by most of the current implementations of the standard.

The graphical representation used in the Function Block Diagram language is shown in Table 5.1.

Table 5.1 Function Block Diagram graphic representation

<table>
<tr><th>Graphical feature</th><th>Semi-graphic form</th><th>Full graphic</th></tr>
<tr><td>Horizontal and vertical lines</td><td>---------------
|
|
|</td><td></td></tr>
<tr><td>Interconnection of horizontal and vertical signal flows</td><td> |
      |
------+--------
      |</td><td></td></tr>
<tr><td>Crossing horizontal and vertical signal flows</td><td>| |
| |
--------|--
| |</td><td></td></tr>
<tr><td>Signal flow corners</td><td>| |
--+   +---
|
--+--+   +---
| | |</td><td></td></tr>
<tr><td>Blocks with connections</td><td>+----+
---| |---
| |
| |---
---| |
+----+</td><td></td></tr>
<tr><td>Connectors</td><td>------->LOAD_JOB>
>LOAD_JOB>-------</td><td>LOAD_JOB>
> LOAD_JOB></td></tr>
</table>

There may be minor differences between the full graphic representation shown in Table 5.1 and the forms used in some implementations of the standard. Full graphic representation is not defined in the standard as the formats of lines, boxes and connections may depend on the resolution and capabilities of the graphics system.

Some implementations may adopt alternative graphic formats because of particular company house styles. The standard describes the main graphical features but not the fine detail. For example, a function block may have right angled or rounded corners, be shown with a shadowed outline or be filled with a contrasting colour or shade.

The standard does not specify the format of line intersections and cross-overs in full graphic implementations. Some means of distinguishing between cross-overs and junctions will be required.

Note: In this book the following conventions have been used in Function Block Diagrams.

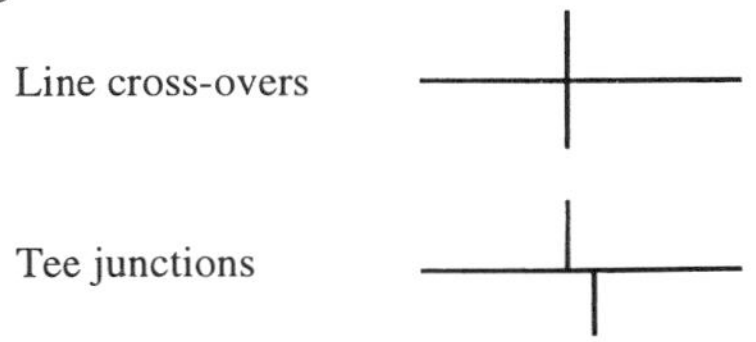

5.2 Methodology

The Function Block Diagram language, FBD, can be used to express the behaviour of programs, functions and function blocks. It can also be used to describe steps, actions and transitions within the Sequential Function Chart language as reviewed in Chapter 8.

Function Block Diagram should be used where the problem involves the flow of signals between control blocks. An FBD network can be regarded as analogous to an electrical circuit diagram where electrical connections depict signal paths between components. Typical uses of FBD include describing control loops and logic. Some program developers who are familiar with electrical circuits or using graphical programming techniques may prefer to use FBD to using a textual language such as Structured Text.

Within FBD certain graphical conventions apply. A function block is depicted as a rectangular block with inputs entering from the left and outputs exiting on the right. Function block type name is always shown within the block, whereas the name of the function block instance is always shown above the block. The formal names of function block inputs and outputs are shown within the block at the appropriate input and output points.

To avoid cluttering diagrams, implementations of the FBD editor on some program stations only depict input and output parameters using their 'pin' numbers. In this case, the full input and output parameter names are accessible by other means, for example by a command from a pull-down menu.

The order in which inputs and outputs are depicted in the function block should always be the same.

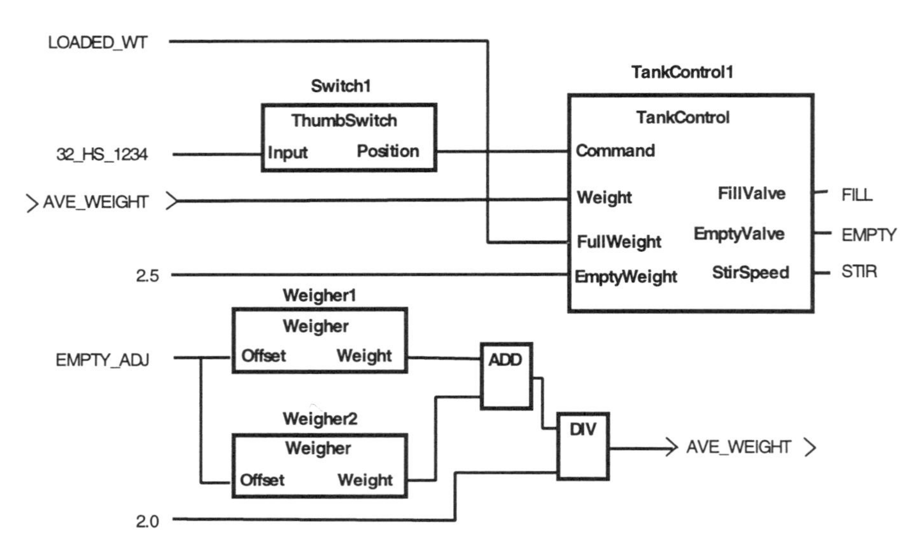

Figure 5.1 Tank control Function Block Diagram example

The main features of the Function Block Diagram language are shown in Figure 5.1 which depicts part of a tank filling program. TankControl1 is the name of an instance of the TankControl function block type. The 'Command' input, which selects the mode of the TankControl function block, is driven by the 'Position' output from function block Switch1; this is an instance of function block type ThumbSwitch. This function block reads the current value from a thumbwheel switch and represents the value at the output 'Position'.

The average weight of the tank is calculated from the function blocks Weigher1 and Weigher2 using ADD and DIV functions - also depicted as rectangular blocks. The signal AVE_WEIGHT is the name of a cross-diagram connector; it is used to aid readability and avoid a graphical connection that would be difficult to draw. It has no significance as far as the program behaviour is concerned. LOADED_WT is the name of a variable supplying the value of the 'FullWeight' input of the TankControl1 block.

Notice that literal constants such as 2.5 can be used as direct input values to functions and function blocks.

Outputs from functions and function blocks can be used to supply values to variables. The outputs FillValve, EmptyValve and StirSpeed of the TankControl1 function block are used to update variables FILL, EMPTY and STIR.

Note: The LOADED_WT signal path has been deliberately placed at the top of the diagram to demonstrate a signal path cross-over. To aid readability, diagrams should be arranged to minimise signal path cross-overs.

For further details on the TankControl function block see Chapter 4, Section 14.

5.3 Signal flow

In a Function Block Diagram network, the signals are considered to flow from the outputs of functions or function blocks to the inputs of other functions and function blocks.

The outputs of function blocks are updated as a result of function block evaluations. Changes of signal states and values therefore naturally propagate from left to right across the FBD network.

Note: The standard implies that signal flows that represent different data types are all depicted using continuous black lines of the same thickness. However, some program station implementations may choose to represent signal flows of particular data types using different colours and line styles.

Negation of boolean signals

When using boolean signals, negated inputs of functions and function blocks can be shown using a small circle at the input point. Similarly, negated outputs can be shown by placing a small circle at the output.

Implementations that do not support this feature may use the NOT function for signal value negation.

In Figure 5.2, the upper and lower FBD networks are functionally equivalent.

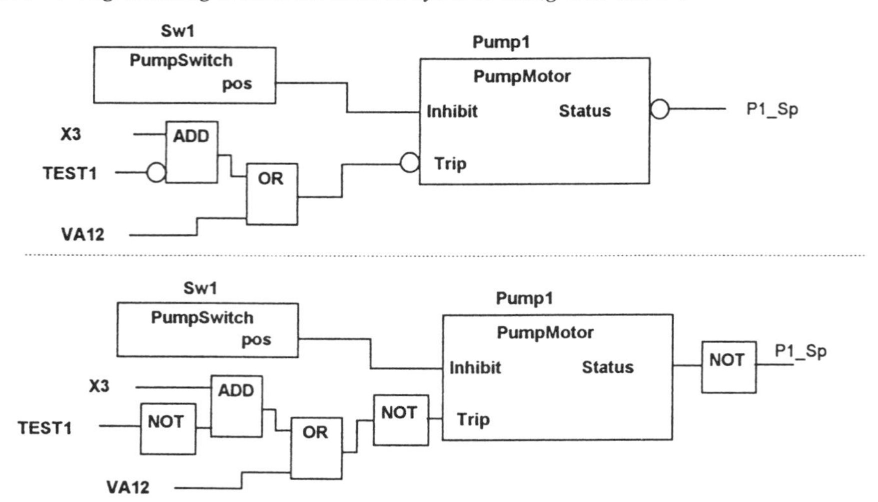

Figure 5.2 Negating boolean signals example

Feedback paths

The standard allows signal paths to be fed back from function block outputs to inputs of blocks on the left of the network, i.e. to the inputs of preceding blocks, as shown in Figure 5.3.

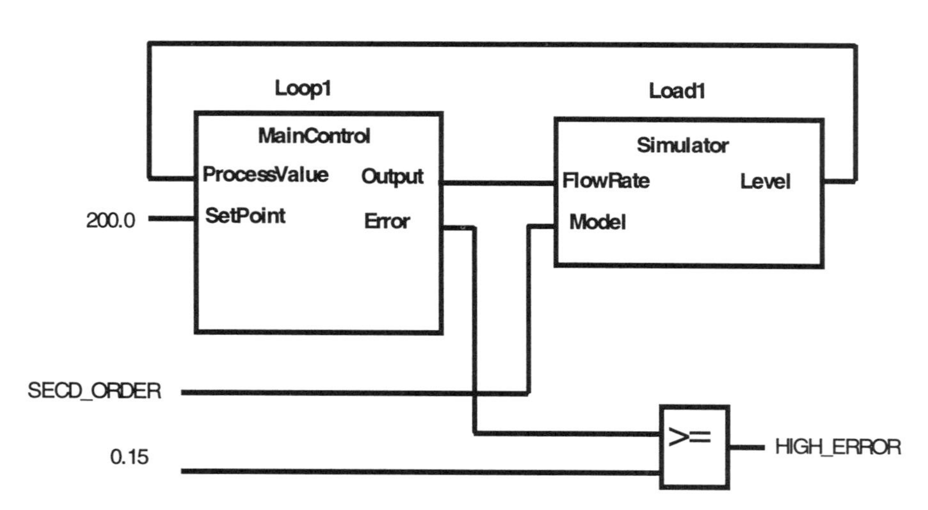

Figure 5.3 FBD feedback example

In Figure 5.3, an explicit feedback path is shown where the output Level from function block Load1 is fed back to the input ProcessValue of the preceding block Loop1.

Implicit feedback paths can also be created by using connectors that link the output parameters from blocks on the right of the network to inputs of preceding blocks.

A feedback path implies that a value within the feedback path is retained after the FBD network is evaluated and used as the starting value on the next network evaluation. However there is no provision in IEC 1131-3 to designate the point in the feedback path where it is broken between network evaluations.

Consider Figure 5.3: there are two ways the feedback path can be evaluated. If function block Loop1 is evaluated before Load1, then at the start of each network evaluation, input ProcessValue will be set to the value of Load1.Level as retained from the previous evaluation. Alternatively, if Load1 is evaluated before Loop1, then input FlowRate will be set to the value of Loop1.Output as retained from the previous evaluation.

As the same FBD network can be interpreted by a PLC system to imply either order of evaluation, the behaviour of networks with feedback paths may vary subtly between different implementations.

Note: When an FBD network is initialised, the first function block input parameter in a feedback path will be initialised using the input's initial value as defined within the function block type definition.

The standard states that implementations may provide a facility so that the order in which function blocks are evaluated in a network can be defined explicitly, for example by giving the function block evaluation order in a list. However, the technique to be used is not defined.

5.4 Network layout

Function Block Diagram networks should be arranged so that the main signal flows can be traced easily. Where possible blocks should be placed to reduce signal path cross-overs and unnecessary changes in signal path direction.

FBD network limitations

The standard does not define specific limitations on the size or complexity of an FBD network. However, it does allow an implementation to limit the number of function block type specifications and the number of function block instances that can be supported in one configuration. Such limits may therefore affect the complexity and size of FBD networks that can be produced when using certain PLC products.

5.5 Function execution control

By default the execution control of functions in an FBD network is implicit from the position of the function in the network. For example, in Figure 5.3, the comparison function '>=' is evaluated after function block Loop1.

However, the control of functions can be made explicit by using the function enable input EN. The EN input is a boolean variable that allows a function to be selectively evaluated. When the EN input is TRUE the function will be evaluated. However, while the value of EN remains FALSE the function will not be evaluated and its output value will not be generated.

The function output ENO is a boolean variable that indicates that the function has completed its evaluation by the ENO value changing from FALSE to TRUE, i.e. a rising edge. By chaining functions together using EN and ENO, it is possible to trigger certain values to be generated when particular conditions arise.

Consider Figure 5.4: as long as either boolean variable add_Acid or pH_High is false, the assignment ':=' and ADD functions are not evaluated. When add_Acid and pH_High are both true, the EN input of the assignment function is asserted TRUE. This causes the assignment function to be evaluated and move the value 100.0 to the variable StirRate. When the function evaluation is complete, the output ENO is asserted which in turn causes the ADD function to be evaluated.

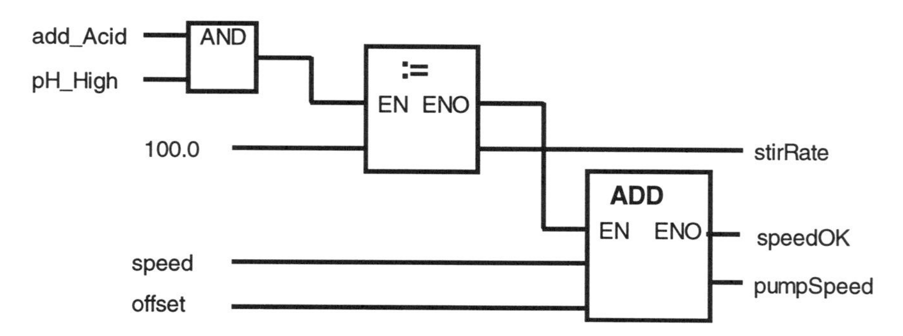

Figure 5.4 Explicit function execution example

Note: The standard is unclear on the use of output values from functions that have EN set false.

Note: A function that has an internal error, such as an arithmetic overflow, will not assert ENO when the evaluation is complete. The value of ENO can therefore also be used to indicate when the result of a function is valid.

5.6 Jumps and labels

It is possible to transfer control from one part of an FBD network to another using a graphical 'jump' facility. Sections of FBD networks can be identified using 'labels'. By assigning a boolean signal to a label identifier, it is possible to transfer control to another part of the network, i.e. associated with the label.

Boolean signals that cause a jump in network evaluation order should be followed by a double arrow and the label identifier.

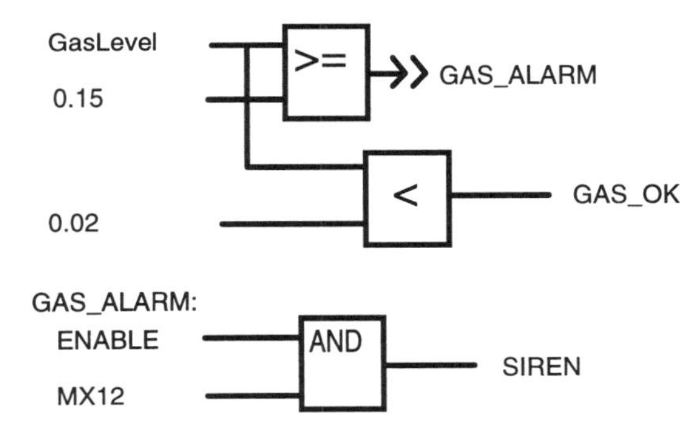

Figure 5.5 FBD jump example

In Figure 5.5, when the value of GasLevel exceeds 0.15, control is transferred to the section of the FBD network identified by the label GAS_ALARM.

Note: IEC 1131-3 does not clarify what happens to the rest of a network after a jump has been triggered. One interpretation is that all linked blocks in the network are evaluated before the jump is made. An alternative interpretation is that the network evaluation ceases immediately, and the network identified by the label is started. For this reason, IEC guideline on using IEC 1131-3 now recommends that jumps within the FBD network should not be used.

If selective evaluation of FBD networks is required, then place the different FBD sections within action blocks of a Sequential Function Chart.

5.7 Network evaluation rules

The order in which function blocks and functions within an FBD are evaluated will vary from product to product. The standard does not attempt to define a strict network evaluation order, although, generally, functions and function blocks on the left of the network will be evaluated before those on the right. Some implementations may also evaluate blocks in an order that runs from top to bottom of the diagram.

However, the following rules should always apply and ensure that data transferred between functions and function blocks are consistent.

- 'No element in a network shall be evaluated unless the states of all inputs have been evaluated.' In other words, a function or function block should not be evaluated unless all of the input values that come from other elements are available.
- 'The evaluation of a network element shall not be complete until the states of all of its outputs have been evaluated,' i.e. outputs of a function block should not be considered available until all outputs have all been updated.
- 'The evaluation of a network is not complete until the outputs of all of its elements have been evaluated.' All outputs of all functions and function blocks should be updated before an FBD network is considered to be complete.
- When data are transferred from one FBD network to another, all the values coming from elements in the first network should be produced by the same network evaluation. The second network should not start evaluation until all the values from the first network are available.

The last rule is particularly applicable to the situation where sets of function blocks in an FBD network are running under the control of different tasks. For example, a set of function blocks under the control of a task that executes every 100 milliseconds may produce output values that are transferred to a second set of function blocks under the control of a different task, for example, that runs every 500 milliseconds.

In this situation, the values of outputs transferred from the first set of blocks to the second should all be created by the same 100 millisecond task evaluation.

5.8 Portability between ST and FBD

There is an assumption by many implementors of the IEC 1131-3 standard that there should be a one to one correspondence between all of the IEC languages. In other words, it should be possible to translate any language into any one of the others. However, it has never been the deliberate intention of the IEC working group that a 100% correspondence should exist. In practice, there is a limited set of core features in each language that can be translated successfully into the other languages. There are also features that are very difficult and, in a few cases, impossible to translate.

Most constructs used in FBD networks can be translated into Structured Text. ST can easily represent the invocation of functions and function blocks and the associated passing of parameter values. For example, the following Structured Text statements are equivalent to the FBD network given in Figure 5.3:

```
VAR
  MainControl : Loop1;
  Simulator : Load1;
END_VAR

(* Invoke the function blocks with *)
(* parameter connections *)
Loop1(ProcessValue := Load1.Level,
      SetPoint := 200.0 );
Load1(FlowRate := Loop1.Output,
       Model := SECD_ORDER);
HIGH_ERROR := Loop1.Output >= 0.15;
```

Note: A number of functions, such as the greater or equal comparison '>=', can be represented in Structured Text as either operators or functions.

Functions that have explicit execution control as discussed in Section 5.5 cannot be represented in Structured Text; there is no syntax for handling functions with the second output ENO!

Translation of Structured Text into FBD networks is even more problematic. Constructs such as IF...THEN, CASE, FOR, WHILE, REPEAT cannot be directly represented graphically except with some ingenuity. It is also not possible to reference elements in an array or structure. For example, the following ST statements have no FBD equivalent:

```
FOR I := 1 TO 100 DO
  rate[I] := 100;
END_FOR;
```

Structured Text should be regarded as the language with the feature super-set. Translation from FBD to ST is always possible provided that explicit function execution control is not used. Translation in the other direction from ST to FBD can only apply to a subset of the ST constructs.

5.9 FBD function block example

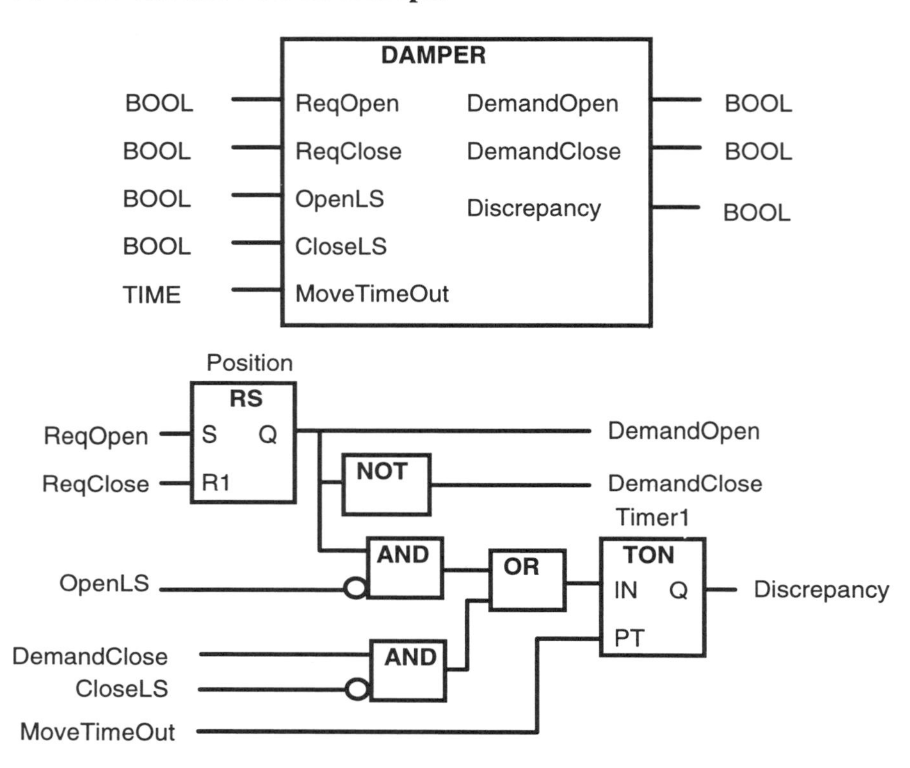

Figure 5.6 Damper function block example

Figure 5.6 is an example of a Function Block Diagram being used to describe the behaviour of a function block for controlling an air damper. Such an air damper might, for example, be used to control the air flow into a furnace or the air for ventilating a region of plant.

The signals ReqOpen and ReqClose signal the requested position for the damper. The requested position is held in an RS bistable function block. The output from the 'Position' bistable and the inverted output are used to produce the signals DemandOpen and DemandClose. These signals drive actuators that move the air damper into the requested position.

Limit switches on the damper signal back the actual position of the damper on the input signals, OpenLS and CloseLS. The limit switch signals are compared with the demanded position using the AND functions. If either check fails, e.g. the position Open is demanded but the limit switch signal OpenLS is not true, the 'on delay' timer Timer1 is started.

If the damper fails to move to the requested position and assert the required limit switch within the time given by the MoveTimeOut input, Timer1 will raise the Discrepancy signal.

Note: The RS bistable and TON timer are IEC standard function blocks as described in Chapter 9.

5.10 Continuous function chart

There is a special form of function block diagram which is used to depict the top level structure of a resource and the assignment of programs and function blocks to tasks: see Chapter 3, Section 42. Although this form of FBD is not given a particular name in the standard, it has become commonly known as a 'Continuous function chart'.

The main difference between a Continuous function chart and a Function Block Diagram is that it also shows the resource and task assignments. Each function block is depicted with the name of the task which controls its execution, as shown in Figure 5.7. Programs may also have the name of their task depicted in a similar manner in a text box at the bottom of the program definition.

If a function block within a program executes under the same task as its parent program, the task association is said to be implicit. In this case, the task name does not need to be shown in the function block graphic.

The format of Continuous function charts is shown in Figure 5.7. Many of the smaller details, such as showing the variables that are assigned to function block inputs, have not been shown to aid clarity.

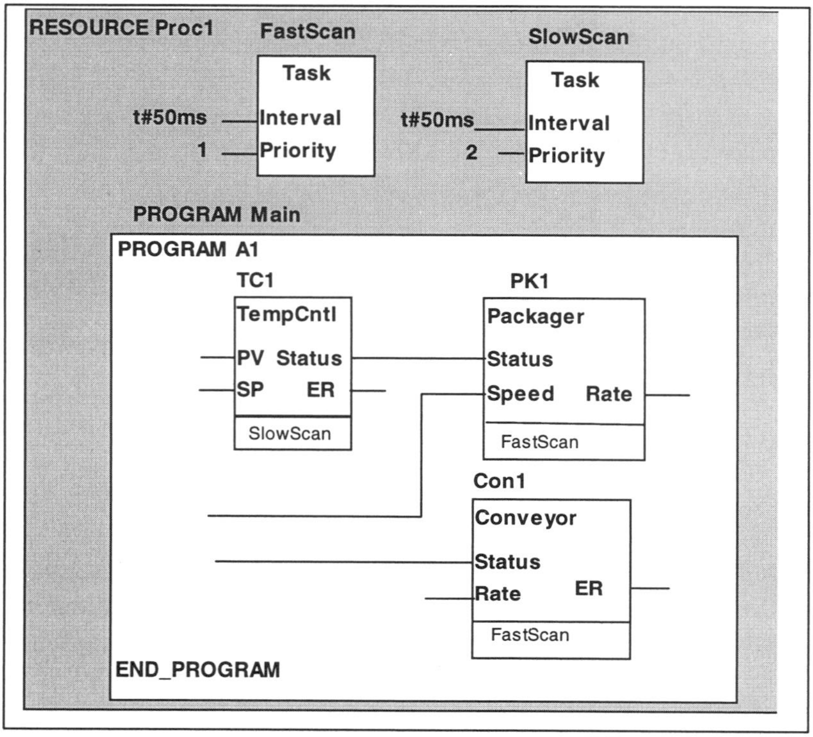

Figure 5.7 Continuous function chart example

Summary

Function Block Diagram is an ideal language to express control behaviour where the solution can be considered as a network of software blocks. FBD is very similar to an electrical circuit diagram, where the lines between the blocks depict signal flows.

- FBD can be used in a wide range of problems including boolean logic and 'closed loop' control.
- Although the standard provides a mechanism through the 'jump' construct for controlling the execution of sections of an FDB network, FBD is only really suitable for behaviour that is continuous. For problems that involve sequencing of some form, the behaviour should be expressed using the Sequential Function Chart (SFC) language as discussed in Chapter 8.

- There is not always a one-to-one correspondence between Function Block Diagram and the other languages. FBD networks can generally be translated into Structured Text; translation from ST to FBD is not always straightforward and, in some cases, is not possible.
- The FBD language can also be used to show the top level structure of resources and programs in a form called Continuous function charts.

Chapter 6

Ladder Diagram

In this chapter we will review Ladder Diagram, the second graphical language of the IEC languages in the set:

> Structured Text, Function Block Diagram, Ladder Diagram, Instruction List and Sequential Function Chart.

Ladder Diagram can be used to express the behaviour of functions, function blocks and programs, and also actions and transitions in Sequential Function Charts.

In this chapter we will review:

- The graphical symbols and conventions used to describe Ladder Diagrams;
- How Ladder Diagram is based on a concept of 'power flow' between contacts which can control the operation of notional 'coils' that represent variables;
- How to develop complex boolean logic expressions;
- How to control functions and function blocks through using connections on a Ladder Diagram 'rung';
- How to control ladder rung execution using 'jumps' and 'labels';
- Restrictions on the portability of code between ST, FBD and LD.

6.1 LD background

Ladder Diagram has been developed by the IEC by considering the most commonly used symbols and terminology used in mainstream PLCs. To anyone who is already familiar with writing programs using 'relay ladder', the symbols defined in IEC 1131-3 will not bring any unwelcome surprises.

If you are not familiar with this form of graphical programming, it is recommended that you read the introduction given in Chapter 1 where some of the main features and limitations of the language are discussed.

Ladder Diagram is based on a technique used to design logic using relays. A Ladder Diagram always has a left hand vertical power rail that notionally supplies power through contacts spread out along horizontal rungs.

Each contact represents the state of a boolean variable. For example, a contact may depict the state of a limit switch or heat sensor. When all contacts in a horizontal rung are made, i.e. in the 'true' state, power can flow along the rail and operate a coil on the right of the rung. Contacts which are normally open present a 'true' state when the contacts are closed. Conversely contacts that are normally held closed, present a 'true' state when the contacts are opened. By arranging alternative paths through a network of contacts it is possible to describe various logic conditions.

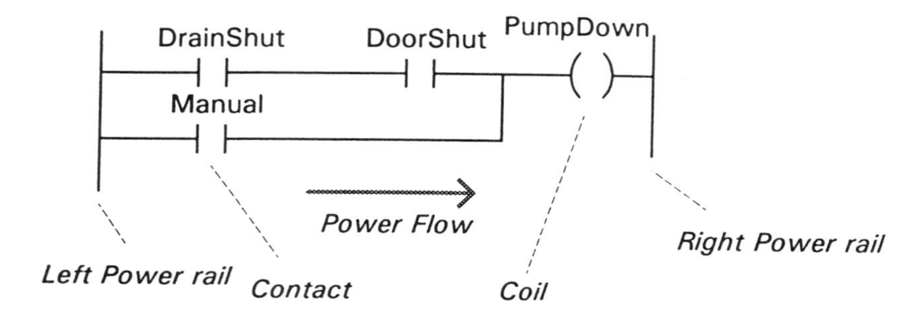

Figure 6.1 Ladder Diagram concepts

Figure 6.1 depicts the main features of the Ladder Diagram language. In this example, PumpDown represents a relay coil which is activated when power is able to flow from the left hand power rail. PumpDown is actually a boolean variable which is set to a true state either when contacts DrainShut and DoorShut are ON, or when contact Manual is ON. The contacts represent the states of boolean variables.

Using the Structured Text language, the Ladder Diagram in Figure 6.1 is equivalent to:

```
PumpDown := (DrainShut AND DoorShut) OR Manual;
```

In Ladder Diagram a contact is assumed to have states ON and OFF (for a normally open contact these states are sometimes referred to as CLOSED and

OPEN) which are equivalent to boolean states 1 (true) and 0 (false). A contact always represents the state of the associated boolean variable. A contact provides a 'read only' access to the state of the boolean variable; it cannot be used to change the value of the variable.

On the other hand, a coil provides 'write only' access and can only be used to update the state of the associated boolean variable when notional power flow can occur via contacts from the left hand power rail.

The standard allows the use of the right hand power rail to be optional. Ladder Diagrams are still quite understandable if the right power rail is not shown.

The network that is associated with the logic for one coil is called a 'ladder rung'.

6.2 LD graphical symbols

The graphical representation used in the Ladder Diagram language is shown in Table 6.1. There may be minor differences between the full graphic representation shown in Table 6.1 and the forms used in some implementations of the standard.

Table 6.1 Ladder Diagram graphic representation

Graphical feature	Semi-graphic form	Full graphic
Horizontal link (power flow)	`----------------`	────────

Interconnection of horizontal and vertical power flows	`      |`
	`      |`
	`------+--------`
	` -----+`
	`      |`	

Left hand power rail connection	`|`
	`|`
	`+-----------`
	`|`	

Right hand power rail connection	`             |`
	`-------------+`
	`             |`	

Connectors (same as for FBD)	------->LOAD_JOB> >LOAD_JOB>-------	──────> LOAD_JOB> > LOAD_JOB>──────

Full graphic representation is not defined in the standard for Ladder Diagram because the formats of lines, boxes and connections may depend on the resolution and capabilities of the graphics provided by the programming system.

The symbols for different types of contacts and coils are shown in Tables 6.2, 6.3 and 6.4. Although not shown in these tables, the name of the associated variable is always displayed directly above the symbol as shown in Figure 6.1.

A range of different contact types is provided to detect normal and inverse states, and rising and falling edges of power flow of the associated boolean variable.

Table 6.2 Ladder Diagram contact symbols

Graphical feature	Semi-graphic form	Full graphic
Normally open contact Normally open contact where power flow occurs from left to right when the associated variable is 1.	`-----| |----`	——| |——
Normally closed contact Normally closed contact where power flow occurs from left to right when the associated variable is 0.	`-----|/|----`	——|/|——
Positive transition-sensing contact Positive rising edge transition contact where power flow occurs for one Ladder Diagram evaluation when the associated variable changes from 0 to 1.	`-----|P|----`	——|**P**|——
Negative transition-sensing contact Negative falling edge transition contact where power flow occurs for one Ladder Diagram evaluation when the associated variable changes from 1 to 0.	`-----|N|----`	——|**N**|——

Note: The exclamation mark '!' can be used in place of the vertical bar '|' in the semi-graphical symbols for contacts.

Table 6.3 Ladder Diagram coil symbols

Graphical Feature	Semi-graphic form	Full graphic
Coil The coil is set to the state according to the power flow coming from the left hand link. If power flow is ON, the coil state is set to ON.	`-----( )----`	—( )—
Negated coil The negated coil is set to the opposite state to the power flow coming from the left hand link. If the power flow is ON, the coil state is set to OFF.	`-----(/)----`	—(/)—
SET coil This coil is set to the ON state when there is power flow coming from the left hand link. The coil remains set until it is RESET.	`-----(S)----`	—(S)—
RESET coil This coil is reset to the OFF state when there is power flow coming from the left hand link. The coil remains OFF until it is SET.	`-----(R)----`	—(R)—

Coil symbols allow the associated variable to be set or cleared on the presence of power flow. The SET and RESET coils allow a variable to be latched on and then cleared by a different condition some time later.

A number of special coil symbols are provided as shown in Table 6.4. The retentive coils can be used where the states of variables need to be held through PLC power failure and resumption. This type of coil may be required for holding the state of critical variables such as plant operation modes, positions of interlocks and so on, i.e. variables that need to be in the correct state when the PLC is powered up following power failure.

Variables associated with retentive coils are automatically declared using VAR_RETAIN.

Table 6.4 Ladder Diagram special coil symbols

Graphical feature	Semi-graphic form	Full graphic
Retentive memory coil The same behaviour as the normal coil, except that the state of the associated variable is retained on PLC power fail.	`-----(M)----`	——(M)——
SET retentive memory coil The same behaviour as the SET coil except that the state of the associated variable is retained on PLC power fail.	`-----(SM)---`	——(SM)——
RESET Retentive Memory coil The same behaviour as the RESET coil except that the state of the associated variable is retained on PLC power fail.	`-----(RM)---`	——(RM)——
Positive transition-sensing coil If the power flow on the left hand link changes from OFF to ON, the variable associated with the coil is set ON for one ladder rung evaluation.	`-----(P)----`	——(P)——
Negative transition-sensing coil If the power flow on the left hand link changes from ON to OFF, the variable associated with the coil is set ON for one ladder rung evaluation.	`-----(N)----`	——(N)——

The positive and negative transition-sensing coils are used to detect a change in state in the power flow. With these types of coil, the associated variable is only set ON for one evaluation of the Ladder Diagram.

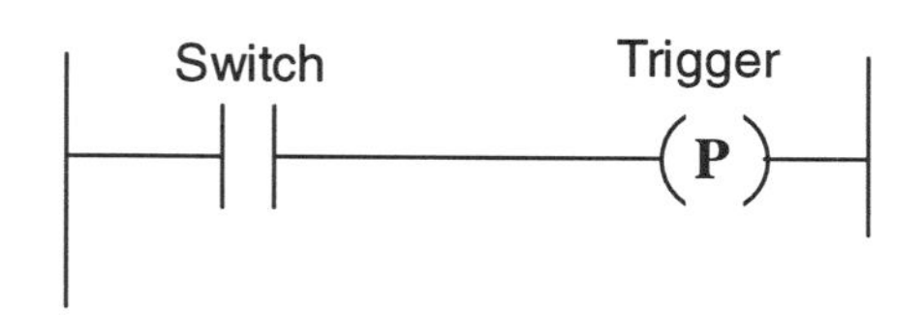

Consider the example of a contact connected directly to a positive transition-sensing coil. The behaviour of the coil over a number of Ladder Diagram evaluations is shown in the following table - notice that the Trigger variable is only set ON once when the Switch changes from OFF to ON.

Evaluation	Switch	Trigger
1	OFF	OFF
2	ON	ON
3	ON	OFF
4	OFF	OFF

6.3 Methodology — relay logic

From the nature of electrical circuits, contacts in a horizontal rung represent logical AND operations. Alternative paths from the main rung provided by vertical paths represent logical OR operations.

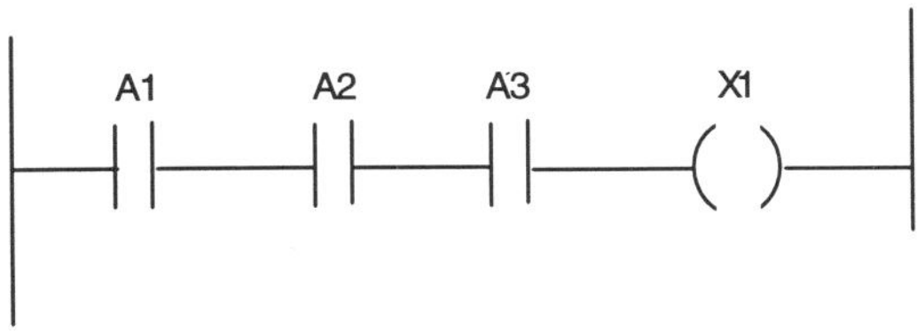

A1 A2 A3 X1

Figure 6.2 Ladder Diagram AND logic example

In Figure 6.2, the Ladder Diagram logic is equivalent to the following Structured Text:

```
X1 := A1 AND A2 AND A3
```

The use of alternative power flow paths via other horizontal rungs connected by vertical paths provides logical OR operations.

In Figure 6.3, the addition of alternative power paths modifies the logic. The Structured Text statement equivalent for this rung is now:

```
X1 := ( A1 OR B1 AND A2 AND A3 ) OR ( C1 AND C2 );
```

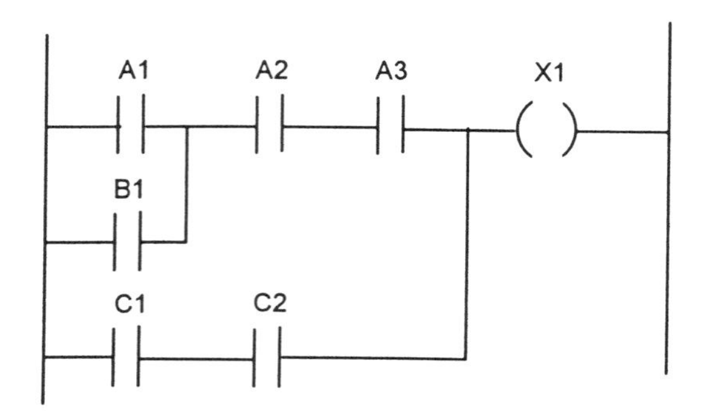

Figure 6.3 Ladder Diagram OR logic example

6.4 Connecting function blocks

Function blocks can be connected into Ladder Diagram rungs provided that they have boolean inputs and outputs. Inputs can be driven directly from Ladder Diagram rungs; outputs can provide power flows for driving coils.

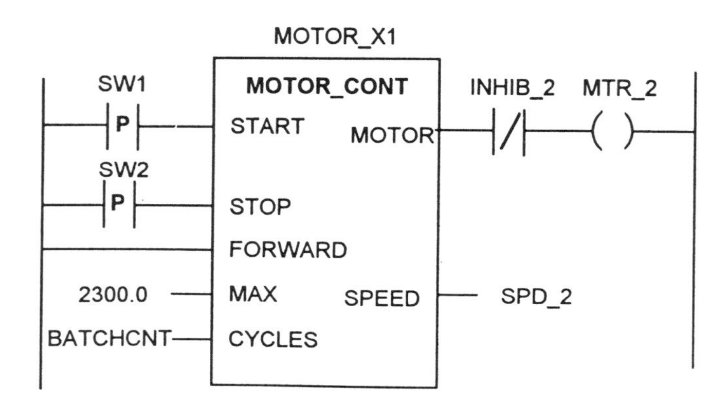

Figure 6.4 Connecting function blocks example

Figure 6.4 is an example of a function block connected into a ladder rung being used to control an electric motor. Notice that boolean inputs START and STOP are driven directly by power flows controlled by contacts. In this case, SW1 and SW2 are boolean variables that represent the states of external push switches. The contacts are positive transition-sensing so that the function block only receives a power flow when the switches are first depressed.

The FORWARD input is connected directly to the left power rail indicating that this input is permanently in the ON (true) state.

Inputs that are not of the boolean data type may be connected to named variables. In this example, the CYCLES input value comes from a variable BATCHCNT. Inputs may also be asssigned values from outputs of other function blocks.

In addition, inputs may be assigned constant values, e.g. a literal value such as 2300.0.

Boolean function block outputs may be connected into horizontal power flows. In this example, variable MTR_2 is only set ON when the function block output MOTOR is ON and the variable INHIB_2 is OFF.

Note: The standard states that each function block needs at least one boolean input connected via a horizontal rung to the left hand power rail in order to be evaluated. It is unclear in the text whether function blocks will be evaluated when all such power flows are OFF due to the states of contacts in the horizontal rungs.
The IEC technical report on the application of IEC 1131-3 states that the evaluation of function blocks can be made explicit using the EN and ENO parameters as discussed in the next section.

6.5 Using functions in LD

Functions may be provided with an additional boolean input EN and an additional output ENO for use in the graphical languages. EN provides a power flow signal into a function. When EN is ON, i.e. 1, the function is enabled and can evaluate its inputs to produce an output. When the function evaluation is complete, the ENO output can be used to provide power flow to other functions or to a coil. The use of EN and ENO is also discussed in Chapter 3, Section 28.

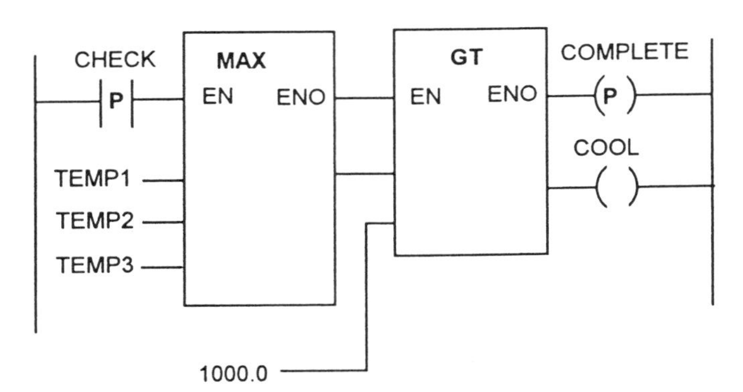

Figure 6.5 Using EN and ENO example

Figure 6.5 depicts an example of using EN and ENO to control the evaluation of functions. Both functions are dormant until the variable CHECK changes from OFF to ON. The positive transition-sensing contact causes power to flow into function MAX for one evaluation. The MAX function delivers the maximum value of variables TEMP1, TEMP2 and TEMP3 at its output and at the same time sets it's ENO output to ON.

This, in turn, allows the 'greater than' (GT) comparison function to be evaluated. It compares the value from the MAX function with 1000.0. If the value is greater, the GT function sets its output ON that changes the state of variable COOL to ON; otherwise COOL remains OFF. In either case, when the comparison is complete, the ENO is set ON, which sets variable COMPLETE to ON, indicating that the temperature check has been completed.

An amendment to the 1993 edition of IEC 1131-3 for the second edition proposes that function blocks may also have the EN and ENO execution control input and outputs.

6.6 Feedback paths

It is possible to create Ladder Diagram rungs which contain feedback loops, i.e. where one or more values used in contacts, and as function and function block inputs, come from variables that are updated when the rungs are evaluated.

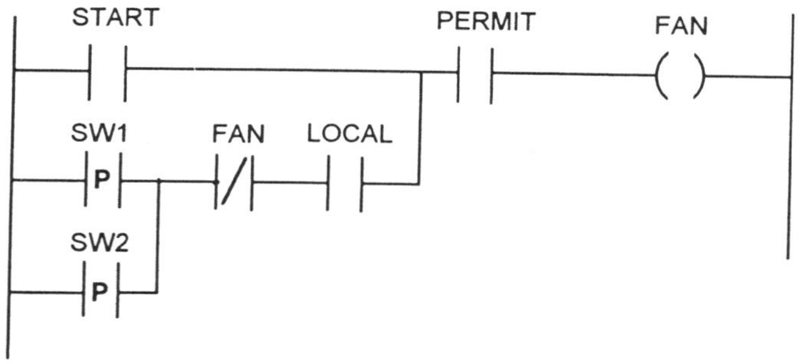

Figure 6.6 LD feedback example

An example of a feedback path in Ladder Diagram is shown in Figure 6.6. Unlike the use of feedback in Function Block Diagram, in Ladder Diagram there is no ambiguity about where the feedback path is broken between evaluations. In Ladder Diagram, all external input values such as those associated with contacts are gathered before each rung is evaluated.

In this example, where the rung uses the state of its own output, the value of FAN associated with a contact is always the value resulting from the previous evaluation.

However, if the value of FAN is used in any following rungs, the latest evaluated state is used.

6.7 Jumps and labels

Using a graphical jump it is possible to transfer control from one part of an LD network to another identified by a label. Assigning a power flow to a label identifier transfers control to another part of the network that is associated with the label.

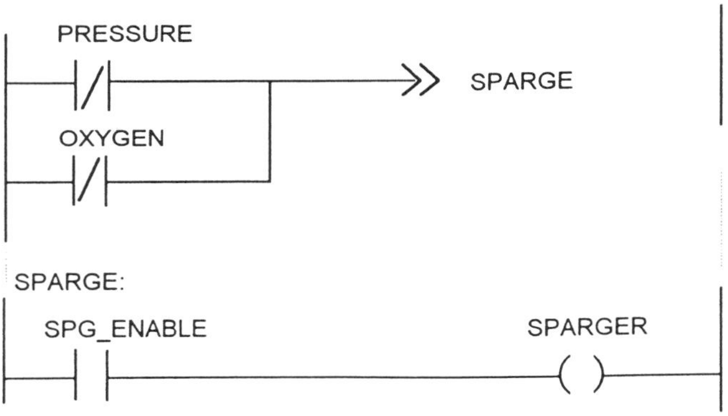

PRESSURE

OXYGEN

SPARGE

SPARGE:

SPG_ENABLE

SPARGER

Figure 6.7 LD jump example

In Figure 6.7, when variables OXYGEN or PRESSURE are OFF, control is transferred to the section of the LD network identified by the label SPARGE. The transfer cannot occur until the network section containing the jump has been fully evaluated.

Note: IEC 1131-3 does not clarify what happens to the rest of an LD network after a jump has been actioned. It is assumed that the ladder rungs following a jump are not evaluated.
For this reason, the IEC guidelines on using IEC 1131-3 now recommend that jumps within an LD network should not be used.
If selective evaluation of an LD network is required, then place the different LD sections within action blocks of a Sequential Function Chart.

6.8 Network evaluation rules

The order in which rungs of a Ladder Diagram network are evaluated is specified in the standard. PLC ladder rungs are always scanned from the top of a Ladder Diagram to the bottom.

This follows the normal convention; PLCs scan ladder programs from the first rung through to the last; the rungs are evaluated in the same order as drawn.

The following rules, that also apply to Function Block Diagram networks, ensure that the result of an LD network evaluation is unambiguous and consistent:

1. 'No element in a network shall be evaluated unless the states of all inputs have been evaluated.' In other words, a ladder rung, a function or function block should not be evaluated unless all of the input values that come from other elements are available.
2. 'The evaluation of a network element shall not be complete until the states of all of its outputs have been evaluated,' i.e. outputs of a function block should not be considered available until all outputs have all been updated; a coil should not change state until all paths in the rung have been evaluated.
3. 'The evaluation of a network is not complete until the outputs of all of its elements have been evaluated.' All outputs of all functions, function blocks and ladder coils should be updated before an LD network is considered to be complete.
4. When data are transferred from one network to another, all the values coming from elements in the first network should be produced by the same network evaluation. The second network should not start evaluation until all the values from the first network are available.

Note: There are situations, particularly with networks that involve feedback, where the application of rule 1 may not be possible. For example, in a network involving feedback it is not possible to provide the initial value of feedback variables prior to evaluating the network.

6.9 Portability between ST, FBD and LD

Networks described using Ladder Diagram, as with Function Block Diagram, are not always readily translated into the other IEC languages. Simple ladder rungs involving mainly AND and OR logic, such as the example rung shown in Figure 6.3, can certainly be expressed in Structured Text and in most cases they can also be expressed using FBD. The following FBD network is equivalent to the rung shown in Figure 6.3.

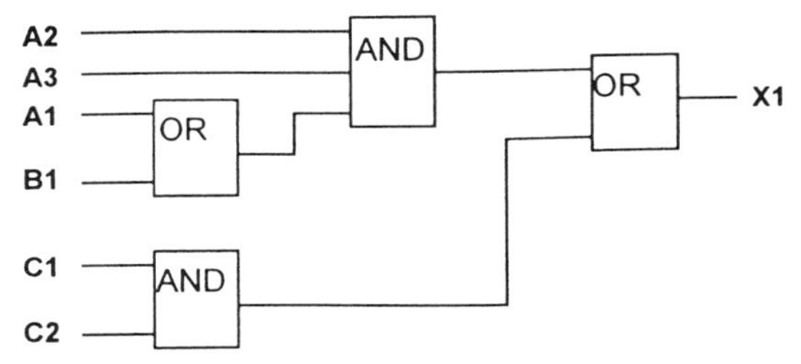

Figure 6.8 FBD logic equivalent

With the 1993 edition of IEC 1131-3, it is very difficult to translate Ladder Diagram rungs involving functions, where EN and ENO are used to pass power flow, into Structured Text. As each function can only have a single output, there is no construct in the ST language that can represent the behaviour of ENO. With some ingenuity, it might be possible to re-express the behaviour of a ladder rung into Structured Text. However, fully automatic one-to-one translation from LD into ST would not be easy to accomplish.

However, an amendment to the 1993 edition of IEC 1131-3 proposes that functions expressed in Structured Text may have multiple outputs: see Chapter 3, Section 28. It will then be relatively straightforward to translate from LD into ST.

Nevertheless, as FBD also allows EN and ENO to be used for function and function block evaluation control, translation between FBD and LD should also be fairly straightforward. Figure 6.9 depicts the FBD network using EN and ENO that is equivalent to the LD network in the example in Figure 6.5.

Note: The positive transition-sensing contact has been replaced by a rising edge detecting function block (R TRIG). This is one of the standard IEC function blocks that are discussed in Chapter 9.

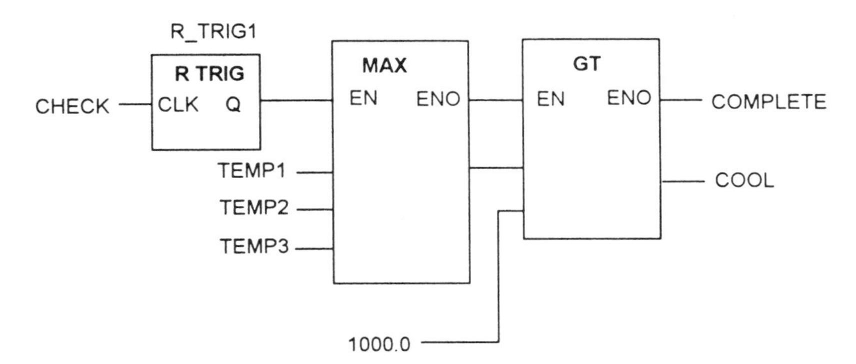

Figure 6.9 FBD equivalent using EN and ENO

Translation from Structured Text into LD networks (as with FBD) is not always possible. Constructs such as IF...THEN, CASE, FOR, WHILE, REPEAT cannot be represented directly in LD rungs. It is also not possible to reference elements in an array or structure. For example, the following ST statements have no LD equivalent:

```
FOR I := 1 TO 100 DO
  rate[I] := 100;
END_FOR;
```

Apart from simple logic expressions, direct language translation between Structured Text and Ladder Diagram is not always possible.

6.10 Ladder Diagram layout

Ladder Diagram can be very effective at describing digital logic in a form that is easy to program and understand. However, use of LD for applications such as arithmetic computations, closed loop control, and sequencing can result in a large number of complex rungs that can be very difficult to program and maintain.

It is therefore recommended that sections of LD be kept fairly short. Where applications are complex, the problem should be broken down into a number of well-partitioned sections. Each section can then be programmed, say, as a function block or function. Ideally LD should be used to describe program organisation units that deal primarily with combinational logic.

The tendency of some programmers, who are very familiar with programming in ladder logic, to write large sections of a program as a long series of complex ladder rungs should be avoided. Such an approach will lose many of the benefits of improved software structure that should be inherent with IEC 1131-3.

When developing Ladder Diagram networks it is worth reviewing the standard set of functions and function blocks. In many cases, a Ladder Diagram network can be simplified by using, say, a built-in function.

LD limitations

There are no specific limitations in the standard that only apply to Ladder Diagram networks. For example, there are no limits on the number of contacts or vertical and horizontal paths that can exist as part of a single LD rung.

However, the standard does allow for product dependent limits on the number of function block type specifications, the number of function block instances and the number of functions that can be supported in one configuration. Such limits may therefore affect the complexity and size of LD networks that can be produced when using certain PLC products.

In practice, the programming system for a PLC will probably have a limit on the width of LD rungs and therefore the number of contacts that can be used in one rung. There will also generally be a limit on the number of rungs used in, say, a definition of a function or a function block.

6.11 LD program example

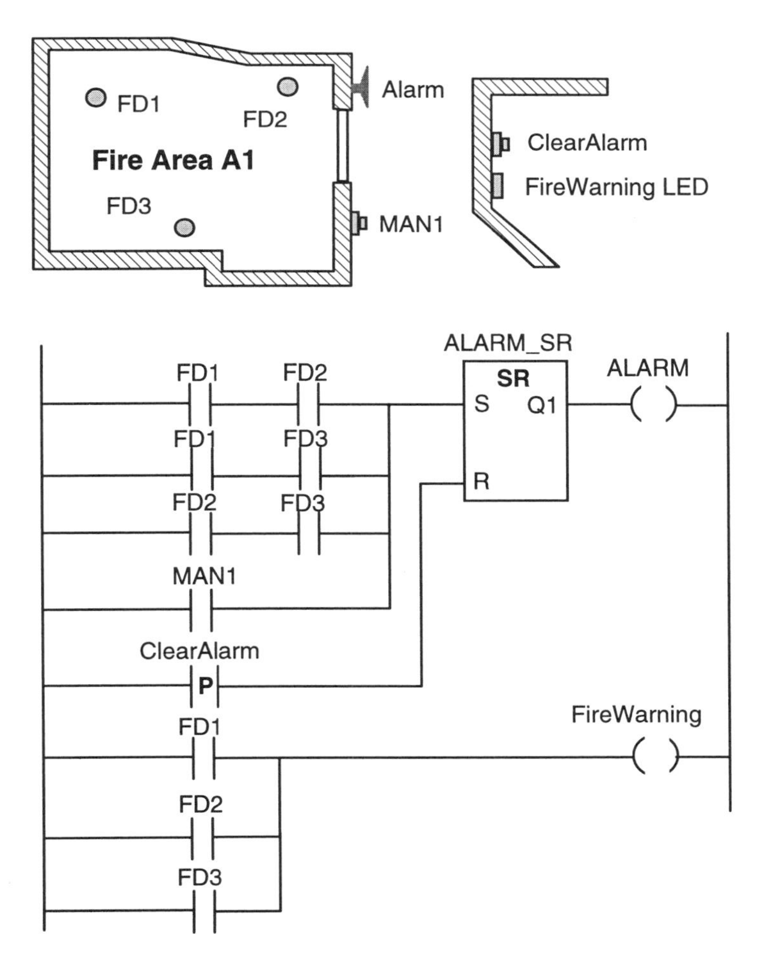

Figure 6.10 Ladder Diagram fire alarm example

Figure 6.10 is an example of a part of a fire alarm system programmed using Ladder Diagram. A fire area is monitored by three fire detectors FD1, FD2 and FD3. There is also a manual push button MAN1 which can be used to trigger the fire alarm. Certain fire detectors can be unreliable and can erroneously signal the presence of fire. To prevent false alarms, the system only triggers the alarm if two or more fire detectors are tripped, i.e. the system is required to perform 2 out of 3 voting.

This is achieved by simply checking for the various combinations of detectors. If any two are ON, then the Alarm_SR function block, a standard RS bistable, is set ON. This in turn drives the Alarm coil.

The Alarm can only be cleared by pushing the ClearAlarm button. The positive transition-sensing contact ensures that if the button remains depressed, a new alarm can still occur. With an SR bistable, the Set signal has a priority over the Reset. This ensures that the Alarm will continue to be ON if the ClearAlarm signal occurs while two or more detectors are ON.

If any fire detector is ON, a Fire Warning LED indicator is illuminated. If this remains lit after the alarm is cleared, this can indicate either that a detector is faulty or that there is a localised fire near a detector.

Note: This example is used purely to demonstrate the LD language and does not represent all the functionality required for an effective fire alarm system.

Note: The RS bistable is one of the IEC standard function blocks as described in Chapter 9.

Summary

Ladder Diagram is suitable for expressing the behaviour of parts of programs that are primarily concerned with combinational logic.

- The language is based on relay logic diagrams, is intuitive and is easy to learn. It is particularly suitable for logic involving complex AND and OR operations.
- Facilities are provided to allow function blocks and functions to be called from Ladder Diagram rungs.
- Although IEC 1131-3 provides a mechanism through the 'jump' construct for controlling the execution of sections of an LD network, the use of jumps should be avoided. It is recommended that all sequencing and selective execution of LD is performed using the Sequential Function Chart (SFC) language as discussed in Chapter 8.

- Ladder Diagram does not always have a one-to-one correspondence with languages such as Function Block Diagram and Structured Text. LD networks containing function calls using EN and ENO cannot be readily converted into Structured Text. However, it is fairly straightforward to translate between LD and FBD.

Chapter 7

Instruction List

In this chapter we will review Instruction List; this is the fourth language of the IEC languages in the set:

Structured Text, Function Block Diagram, Ladder Diagram, Instruction List and Sequential Function Chart.

Instruction List is a low level language that can be used to express the behaviour of functions, function blocks and programs, and also actions and transitions in Sequential Function Charts.

In this chapter we will review:

- The basic structure of the Instruction List language;
- The behaviour of standard operators;
- How to control the flow of program execution using conditional operators, jumps and labels;
- How to call function blocks and functions in IL;
- Guidelines on writing IL;
- Restrictions on the portability of code between IL and the other IEC languages.

7.1 IL background

Instruction List is a low level language which has a structure similar to a simple machine assembler. The IEC has developed Instruction List by reviewing the many low level languages offered by a wide range of PLC manufacturers. The IL language, as defined in IEC 1131-3, provides a range of operators that represent those most commonly found in proprietary instruction list languages of current day PLCs.

The basic structure of Instruction List is very simple and easy to learn. It is ideal for solving small straightforward problems where there are few decision points and where there are a limited number of changes in program execution flow.

A number of the manufacturers offering IEC 1131-3 based PLCs have chosen to support IL in preference to Structured Text. For a PLC designer, one of the main advantages of IL over ST is that IL is far easier to implement. It is fairly straightforward to develop a PLC that can interpret IL instructions directly. With some systems it is possible to download IL programs to a PLC without going through a program compile and build process. In contrast, Structured Text generally has to be compiled to the native micro-processor assembler of the PLC, such as an 80486 assembler, before it can be down-loaded and run.

Instruction List is sometimes regarded as the language into which all the other IEC languages can be translated. It is the base language of an IEC compliant PLC, into which all other languages such as Structured Text and Function Block Diagram can be converted. However, there are implementors who have taken a different view, and tend to regard IL as a language strictly for small PLCs and treat Structured Text as the base language.

It should be emphasised that the IEC 1131-3 standard does not imply that any particular language should be treated as the base language. Instruction List is merely another language which can be used, if preferred, for solving certain control problems.

As is found when using any machine assembler language, there are a number of disadvantages to using Instruction List over using a high level language such as Structured Text. For example, it is not possible to check the structure of a section of an IL program with the same rigour as a section written in ST. It is also harder to follow the program flow in IL than in ST.

Perhaps one advantage of IL is that, in some circumstances, it may be used to write tight, optimised code for performance critical sections of a program.

7.2 IL language structure

Instruction List is a language which consists of a series of instructions where each instruction is on a new line. An instruction consists of an operator followed by one or more operands. An operand is the 'subject' of the operator. Operands represent variables or constants as defined in Chapter 3, 'Common elements'. A few operators can take a series of operands, in which case each operand should be separated by a comma.

Each instruction may either use or change the value stored in a single register. The standard refers to this register as the 'result' of an instruction. The 'result' can be overwritten with a new value, modified or stored in a variable. The result register is sometimes alternatively referred to as the accumulator or 'accu'.

IL also provides comparison operators which can compare the value of a variable with the register value, i.e. with the current result. The result of the comparison is written back to the register. Various types of jump instruction are provided that can test the current register value and, if appropriate, jump to a named label. Labels can be used to identify various entry points for jump instructions. Each label should be followed by a colon.

The following figure depicts the main features of the IL language:

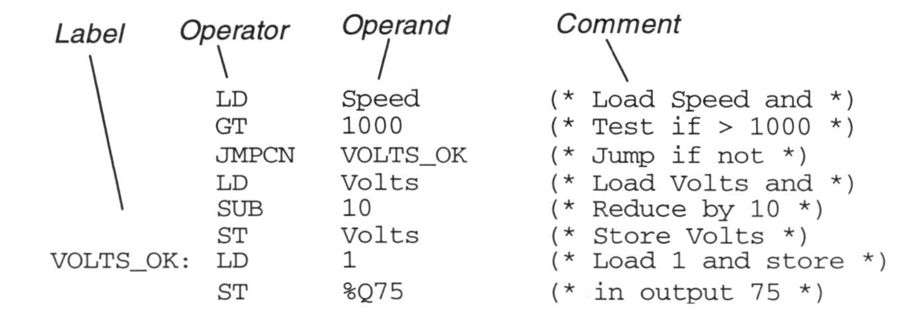

Figure 7.1 IL language structure

The IL instructions in Figure 7.1 are equivalent to the following Structured Text statements:

```
IF Speed > 1000 Then
 Volts := Volts - 10;
END_IF;
%Q75 := 1;
```

Consider the first line of the IL example; LD represents the load operator, and Speed is the operand and identifies the value of an integer variable. The LD instruction loads the current value of variable Speed into the result register.

The 'greater than' GT operator compares the value of the register with literal (constant) 1000. If the value of the register is greater than 1000, the result of the GT operator will be 1 which is equivalent to a boolean true value. Otherwise it will be set to 0, a boolean false value.

It is important to note that the value of Speed is no longer available in the register and that the datatype of the value in the register has changed from integer (INT) to boolean (BOOL).

The JMPNC instruction jumps to the label VOLTS_OK if the result register contains 0, i.e. when the comparison fails.

If the comparison is successful, the jump is not made and the instructions that follow are executed. In this case, the value of variable Volts is loaded into the register. The SUB operator subtracts 10 from the register. The ST operator then stores the result back to variable Volts.

The last two instructions load 1 into the 'result' register and then store the register value into the output 75.

In order to improve readability of the language, IL instructions are structured so that labels, operators and operands appear in fixed tabulation (tab) positions.

7.3 Comments

Instruction List comments have the same format as in Structured Text. However, with IL, a comment can only be placed at the end of the line. It is not acceptable for a comment to appear at the beginning of a line or between an operator and an operand.

One or more blank lines can be inserted between instructions to break up long sequences of instructions and aid readability.

7.4 Instruction semantics

The standard states that the general behaviour or semantics of an IL instruction for the majority of operators can be described by the following expression:

```
NewResult := CurrentResult Operator Operand
```

'CurrentResult' represents the value currently held in the 'result' register prior to the instruction being executed; it is the value left over from the previous instruction.

'NewResult' is the new value to be loaded into the 'result' register as produced by executing the instruction. It is the value produced by the 'Operator' acting on values of 'CurrentResult' and the 'Operand'. The 'NewResult' becomes the 'CurrentResult' for the next instruction.

For example, consider the instruction 'SUB 10' in Figure 7.1 that uses the Subtract operator. This is equivalent to the expression:

```
NewResult := CurrentResult SUB 10
```

The SUB operator takes 10 from the 'CurrentResult' and stores the value in the 'NewResult'.

With comparison operators, the 'CurrentResult' is compared with the operand. The 'NewResult' is set to 1 (true) if the comparison is successful; otherwise it is set to 0 (false).

For example, the behaviour of the comparison instruction 'GT 1000' in Figure 7.1 that uses the 'greater than' operator, is equivalent to:

```
NewResult := CurrentResult GT 1000
```

A few IL operators do not fit this general model. For example, the ST (store) operator can be described by the expression:

```
Operand := CurrentResult
```

This is because the function of the ST operator is to store the current value held in the result register into the location defined by the Operand field.

Tables 7.1 and 7.2 describe the standard IL operators. Additional text and notes are given with the tables to describe the operator behaviour.

Deferred execution

Special modifiers called the parenthesis (or bracket) modifiers allow sections of IL instructions to be deferred, i.e. to produce intermediate results that do not affect the current 'result register'. This has the same effect as the use of brackets in normal arithmetic and boolean expressions.

For example:

```
LD    A      (* Add A to B*)
ADD   B      (* hold the value in result reg. *)
MUL(  A      (* Defer MUL until (A-B) available *)
SUB   B
)            (* Now multiply by (A-B) *)
```

In this example, when the MUL operator is reached, the result of A + B is held in the result register. The left bracket '(' following the MUL operator indicates that the multiply operation will be deferred until the right bracket ')' operator is reached. The value of operand A is then loaded into a temporary result register.

The SUB B instruction produces the value of A - B, which is held in the temporary result register. The final ')' operator takes the value in the temporary result register and multiplies it by the value held in the main result register. This sequence of instructions is equivalent to:

```
result := (A + B) * (A - B);
```

Note: There is no precedence with IL operators. Hence the brackets are needed in this expression, i.e. around (A + B), to indicate that the add operator occurs before the multiplication operator.

The parenthesis operators provide a similar function to a stack. A number of deferred operators can be active at any time allowing fairly complex nested operations to be performed.

Consider the following example:

```
LD     X      (* Load X                 1 *)
ADD(   B      (* Defer ADD, load B      2 *)
MUL(   C      (* Defer MUL, load C      3 *)
ADD    D      (* Add D                  4 *)
)             (* Multiply result        5 *)
)             (* Add result             6 *)
```

This is equivalent to:

```
X + (B * ( C + D))
```

The following table shows how the deferred instructions in the example cause values to be stored in a stack of result registers. The table gives the values in the top three result registers that exist after each instruction in this example has been evaluated.

Instruction	Top result register	Top -1 result register	Top-2 result register
1	X		
2	X	B	
3	X	B	C
4	X	B	C+D
5	X	B*(C+D)	
6	X+B*(C+D)		

Note : The standard does not indicate that the use of jump instructions within parenthesised sections is illegal. Clearly, in practice, jumps from within parenthesised instructions may produce unpredictable results and should be avoided.

Modifiers

Some IL operators can take single letter modifiers after the operator mnemonic that change the semantics of the instruction.

If the operand contains a boolean value, the 'N' modifier can be used to negate its value. With jump operators, the 'N' modifier indicates that the jump will use the negated value of the 'result register'.

The 'C' modifier can only be used with jump operators and indicates that a jump is made conditionally on the boolean value of the result register.

For example:

```
LD     %IX10     (* Load Input 10 *)
ANDN   Switch1   (* AND NOT switch1 *)
JMPNC  Lab1      (* Jump if Not true *)
```

The ANDN operator inverts the value of Switch1 and ANDs the result with the value of input 10 (%IX10). The JMPNC jump operator jumps to label 'Lab1' if the result is not boolean true.

7.5 Operators

The main IL operators are listed in Table 7.1; the comparison operators and instructions concerned with changing the flow of execution control are listed in Table 7.2.

Table 7.1 Main IL operators

Operator	Modifiers	Operand	Comments
LD	N	ANY 1	Load operand into result register
ST	N	ANY 1	Store result register into operand
S	Note 2	BOOL	Set operand true
R	Note 2	BOOL	Reset operand false
AND	N,(	ANY 1	Boolean AND
&	N,(	ANY 1	Boolean AND (equivalent to AND)
OR	N,(	ANY 1	Boolean OR
XOR	N,(	ANY 1	Boolean exclusive OR
NOT		ANY 1	Logical negation (one's complement)
ADD	(	ANY 1	Addition
SUB	(	ANY 1	Subtraction
MUL	(	ANY 1	Multiplication
DIV	(	ANY 1	Division
MOD	(	ANY 1	Modulo-division

Table 7.2 IL comparison and jump operators

Operator	Modifiers	Operand	Comments
GT	(	ANY 1	Comparison greater than
GE	(	ANY 1	Comparison greater than, equal
EQ	(	ANY 1	Comparison equal
NE	(	ANY 1	Comparison not equal
LE	(	ANY 1	Comparison less than, equal
LT	(	ANY 1	Comparison less than
JMP	C,N	LABEL	Jump to label
CAL	C,N	NAME	Call function block, see following sections on the alternative call syntax
RET	C,N		Return from function or function block (see Note 4)
)			Execute the last deferred operator

Note 1: Operators that can take operands of data type ANY are said to be 'overloaded'. That is, the operator can be used with any of the different elementary data types such as integers SINT, INT, time data types such as DATE_AND_TIME, floating point values of type REAL and so on. For a full list of elementary data types see Chapter 3, 'Common Elements'. The operator will only produce a valid result if the current value in the result register is of the same data type as the operand.

Note 2: The Set and Reset operators can only be used with operands of boolean data type.

Note 3: Operators that can have more than one modifier may be used with either or both modifiers. For example there are four forms of the AND operator:

Operator	Semantics
AND	Boolean AND
AND (	Deferred boolean AND
ANDN (	Deferred boolean AND, invert deferred result
ANDN	Invert boolean AND

Note 4: The RET operator is used to return from a function or function block. The modifier C means that the return is made conditionally. If the result register contains boolean true, i.e. 1, the return is made. The N indicates that the result register should be inverted, i.e. false for the return to be made.
With JMP, CAL and RET instructions, the negate 'N' modifier can only be used with the condition modifier 'C', and indicates that the jump is made conditionally on the negated value of the result register, i.e. when the value is boolean 0 (false).

7.6 Calling functions and function blocks

Within Instruction List, the standard provides three different formats of the CAL operator for calling function blocks, and two alternative formats for calling functions. The syntax for calling functions and function blocks is not the same.

Calling a function block using a formal call with an input list

With this format, the CAL operator is followed by a list of function block parameters with their values. Each input parameter must be identified by name. The value of each parameter can be given directly or calculated and passed into the IL current result register (accumulator).

Example:

```
CAL    LOOP1(
SP:= 300.0
PV :=(
LD %IW20
ADD 10
)
)
```

This calls function block instance LOOP1 with parameter SP set to 300.0 and parameter PV set to the value of input word 20 plus 10.

Calling a function block using an informal call

In this format, the values to each input should be set-up prior to calling the function block.

Example:

```
LD    300.0
ST    LOOP1.SP
LD    %IW20
ADD   10
ST    LOOP1.PV
CAL   LOOP1
```

This is equivalent to the previous example; LOOP1 is called with parameter SP set to 300.0 and parameter PV set to the value of input word 20 plus 10.

Calling function blocks using input operators

This format can only be used with some of the IEC standard function blocks as defined in Chapter 9. A number of IL operators are reserved for use with commonly used function blocks such as the SR (Set/Reset) bistable, or the CTU up-counter. A full list of such operators is given in the next section. They are provided as a short-hand form for calling commonly used function blocks.

Example:

```
S1    Latch1
LD    10
PV    CTU1
CU    CTU1
```

The S1 operator can only be used with an SR bistable. In this case the S1 instruction causes Latch1, a function block instance of type SR, to be set.

The PV (preset value) operator can only be used with counter function blocks. In the example, the PV instruction loads 10 from the result register into the PV parameter of the up-counter function block CTU1. The CU instruction calls the up-counter function block with the counter enable parameter set true, i.e. to start counting.

Note: Parameters that are not supplied in a function block call will either take the previous value they were assigned, or take their default value from initialisation.

Calling functions using a formal call

In a formal function call, the function name is given, followed by the values of the input parameters.

Example:

```
SHR(
IN := %IW30
N:= 10
)
```

An amendment to the 1993 edition of IEC 1131-3 for the second edition proposes that when a formal call is made to a function, then optionally the value of the first parameter can be supplied directly from the current result.

Example:

```
LD    %IW30
SHR(
N:= 10
)
ST    %QW100
```

Calling functions using an informal call

In an informal function call, the function name is given, followed by the values of the input parameters. The value of the first parameter is supplied by the current result.

Example:

```
LD   %IW30
SHR  10
ST   %QW100
```

The SHR (shift right) function requires two input parameters. In this example, the first parameter is loaded from the current result, i.e. the contents of input word 30; the second parameter has value 10, i.e. the count of bits for the shift operation. The result from the function is returned as the current result. This is loaded to output word 100.

7.7 Defining functions and function blocks

Instruction List can be used to define the behaviour of functions and function blocks. When used to define a function, the value returned by the function is provided by the last value in the result register (accumulator).

When used to define function blocks, IL can reference the function block inputs (VAR_INPUT) and write values to outputs (VAR_OUTPUTS) that are declared in the function block declaration.

For further details on using IL to define functions see Section 7.12 IL program example'.

7.8 Function block operators

The full list of reserved operators for use with standard function blocks is given in Table 7.3.

Table 7.3 IL Function block operators

Operator	Function block type	Comments
S1,R	SR bistable	Set and Reset the SR bistable.
S,R1	RS bistable	Set and Reset the RS bistable.
CLK	R_Trig, rising edge detector	Clock input to the rising edge detector function block.
CLK	F_Trig, falling edge detector	Clock input to the falling edge detector function block.
CU,R,PV	CTU, up-counter	Control parameters for the CTU up-counter function block, to count-up (CU), reset (R) and load (PV).
CD,PV	CTD, down-counter	Control parameters for the CTD down-counter, to count-down (CD), load LD and set the count minimum (PV).
CU,CD,R, PV	CTUD, up/down-counter	Control parameters; same as for up- and down-counters.
IN,PT	TP, pulse timer	Control parameters for the pulse timer, to start timing (IN) and set-up the pulse time (PT).
IN,PT	TON, on delay timer	Control parameters for the on-delay timer, to start timing (IN) and set-up the delay time (PT).
IN,PT	TOF, off delay timer	Control parameters for the on-delay timer, to start timing (IN) and set-up the off-delay time (PT).

7.9 Deficiencies and proposed extensions

Of all the five IEC languages, the definition of the Instruction List language has been found to be the most contentious. The semantics, i.e. the way the instructions operate, are not always fully described in the standard.

One of the main problems is that the standard does not fully define the behaviour of a conceptual processor, or virtual machine, that can execute IL instructions. For example, it is unclear how the result register stores values of different data types. It is particularly difficult to see how the accumulator can be used to store multi-element variables, such as structured variables, arrays or strings.

The run-time behaviour for various error conditions is also not documented. For example, the behaviour following the use of an operator with an inappropriate data type is not described.

7.10 Portability between IL and other languages

The conversion of sections of Instruction List into the other languages is very difficult. It could be achieved if a restricted range of IL operators were used and the instructions written using a strict format.

Conversion of the other languages into IL is easier but by no means always straightforward.

For example consider the following section of a Function Block Diagram:

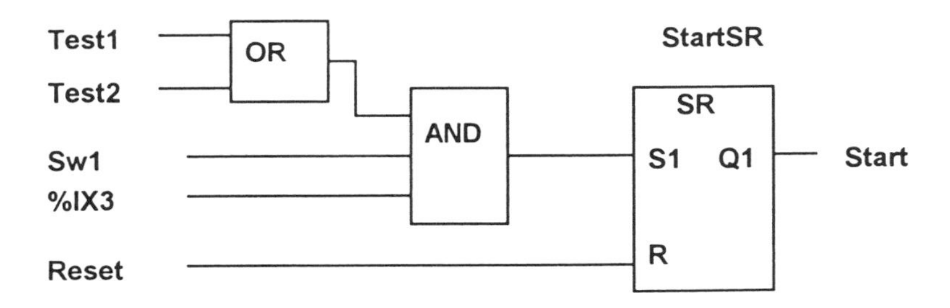

This can be written in Instruction List as follows:

```
LD    Test1        (* Test1 OR *)
OR    Test2        (* Test2 *)
AND   Sw1          (* AND Sw1 *)
AND   %IX3         (* AND input 3 *)
ST    StartSR.S1   (* Set input of StartSR *)
LD    Reset        (* Load value of Reset *)
ST    StartSR.R    (* Store in reset input *)
CAL   StartSR      (* Call fb. StartSR *)
LD    StartSR.Q1   (* Load output Q1 *)
ST    Start        (* and store in Start *)
```

The execution control parameters EN and ENO used in the FBD and LD languages can be used in IL. They are treated as extra parameters of a function or function block.

7.11 Good programming style

Because Instruction List is a fairly low level language, great care should be taken in laying out the code to ensure that it is easy to read and maintain. It is important that IL instructions are well commented. Jump instructions should be used sparingly so that the flow of execution through the code is easy to follow.

Particular care should be taken to ensure that the value in the result register is always appropriate for an IL operator.

7.12 IL program example

In this example we will consider the use of IL for defining a function for calculating the travel distance between two points on a flat surface. Such a function might be required for controlling a numerically controlled (NC) machine that needs to traverse between two drilling points. Assume that the positions of the two points are given by x and y co-ordinates as shown in Figure 7.2.

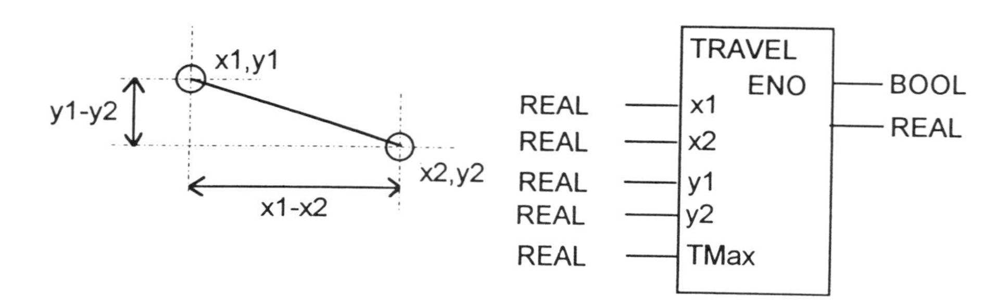

Figure 7.2 IL function example

From trigonometry, the distance between the two points can be expressed in Structured Text as:

```
Travel_distance := SQRT (( X1-X2)*(X1-X2)
                       + (Y1-Y2)*(Y1-Y2));
```

Assume that this function will be used in the graphical languages. The ENO output signal is therefore required for passing execution control on to other functions when the function has completed successfully, see Chapter 5, Section 5.

If the travel distance is less than TMax, the function should set ENO to indicate that it has computed the travel distance successfully. If the travel distance exceeds

TMax, the computation has produced a travel distance outside the range of the controlled machine, in which case the ENO output should not be set.

The full definition of a function TRAVEL( ) to calculate the required movement expressed using Instruction List is as follows:

```
FUNCTION TRAVEL : REAL
 VAR_INPUT
   X1,X2,Y1,Y2 : REAL; (* X and Y Co-ordinates *)
   TMax    : REAL:     (* Maximum travel distance *)
 END_VAR
 VAR
   temp : REAL;   (* Intermediate values *)
 END_VAR

     LD    Y1
     SUB   Y2      (* Subtract Y2 from Y1 *)
     ST    Temp    (* Store Y1-Y2 in Temp *)
     MUL   Temp    (* Multiply by Temp to square *)
     ADD(  X1      (* Defer ADD *)
     SUB   X2      (* Subtract X1 from X2 *)
     ST    Temp    (* Store X1-X2 in Temp *)
     MUL   Temp    (* Multiply by Temp to square *)
     )             (* Add two squared values *)
     CAL   SQRT    (* Call Square root fun. *)
     ST    TRAVEL(* Setup function result*)
     GT    TMax    (* Greater than TMax ? *)
     JMPC  ERR     (* Yes, Jump to Error *)
     S     ENO     (* Set ENO *)
     RET           (* Normal return *)

  ERR:
     RET          (* Error return, ENO not set *)

 END_FUNCTION
```

Note: The value for the SQRT function is supplied from the current result, as discussed in Section 7.6.

Summary

Instruction List is a low level textual language which can be used for describing the behaviour of functions and function blocks, and actions and transitions used in the Sequential Function Charts.

- It is simple to implement and can be directly intepreted within a PLC. It has therefore been adopted by a number of PLC manufacturers for their small to medium sized PLCs.
- It can be used to write tight, optimised code for performance critical operations.
- Operators are provided to manipulate variables of all the elementary data types, call functions and function blocks.
- It is ideal for solving small straightforward problems where there are few decision points and where there are a limited number of changes in program execution flow.
- Care should always be taken when developing programs using IL as there are only limited language validation checks that can be undertaken on a programming station. The majority of language violations, such as data type inconsistencies, can only be detected at run-time.
- Conversion from IL into the other IEC languages is not always possible. However, conversion from the other languages into IL can generally be achieved.

Chapter 8

Sequential Function Chart

This chapter reviews the IEC Sequential Function Chart (SFC) graphical language; this is the final language of the IEC languages in the set:

Structured Text, Function Block Diagram, Ladder Diagram, Instruction List and Sequential Function Chart.

SFC is a powerful graphical technique for describing the sequential behaviour of a control program. This chapter includes:

- The basic concepts of the Sequential Function Chart language;
- The definition of sequences in terms of steps and transitions;
- The use of the other various IEC languages for defining the behaviour of SFC elements, i.e. steps, transitions and actions;
- How to define when actions are executed;
- The use of SFC for top down design;
- Safe and unsafe chart designs.

8.1 SFC background

The definition of the IEC Sequential Function Chart language has been derived from current techniques that are used for depicting sequential behaviour. Many PLC manufacturers have offered some form of graphical language for describing sequences for a number of years. The majority of the European based PLC manufacturers now provide Grafcet, a graphical language based on a French national standard.

Describing the behaviour of a system in terms of states and transitions, depicted as circles and arcs, was originally defined in a methodology called Petri-net. Petri-net is now a well established technique, particularly in computer systems design, for formally describing the behaviour of programs that have multiple states.

Grafcet was evolved from Petri-net as an industrial form of the methodology. In 1988, the IEC published the standard IEC 848, 'Preparation of function charts for control systems'. This defines a graphical language for depicting sequences based very closely on ideas from the French Grafcet standard.

Most of the definitions in IEC 848 have been directly imported into the IEC 1131-3 standard to form the Sequential Function Chart language. SFC therefore has many features that resemble Grafcet and IEC 848. The main enhancements to the methodology have been due to the need to integrate SFC with the other IEC 1131-3 languages.

SFC allows all the sequential aspects of a control program to be described. SFC can be used at the top level to show the main phases of a process, such as 'Start-up', 'Pumping', 'Emptying', or the main states of a machine, like 'Running', 'Stopped' etc. It can also be used at any other level. For example, SFC can be used at a low level to describe the behaviour of a function block handling a serial communications device with various states, such as 'Off', 'Carrier Detected', 'Transmitting' and so on.

One of the most important aspects of SFC is that it shows the main states of a system, all the possible changes of state and the reasons why those changes could occur.

SFC language can be used to partition a control problem so that only those aspects relevant to a particular phase need to be considered and executed. This is very important from a performance point of view because only code concerning active steps is executed.

One or more separate SFC networks can be used to describe the behaviour of programs, function blocks or action blocks. Each separate SFC network has a single initial step which is activated when the associated program organisation unit is initialised.

8.2 Chart structure

A sequence in SFC is depicted as a series of steps shown as rectangular boxes connected by vertical lines as shown in Figure 8.1. Each step represents a particular state of the system being controlled. Each connecting line has a horizontal bar representing a transition. A transition is associated with a condition which, when true, causes the step before the transition to be deactivated and the step that follows the transition to be activated. The flow of control is generally down the page, but branches can be used to go back to earlier steps.

The 'Fill' step remains active while the vessel is being filled with reactants.

When the VesselFull signal is detected, the 'Fill' step is deactivated and the 'Stir' step is started. The 'Stir' step's duration is timed using an elapse timer function block called Timer1. When Timer1's elapsed time reaches 1 hour, the 'Stir' step ceases and the 'Drain' step becomes active.

The final 'Stop' step is reached when the StartSwitch is placed in the 'Off' position.

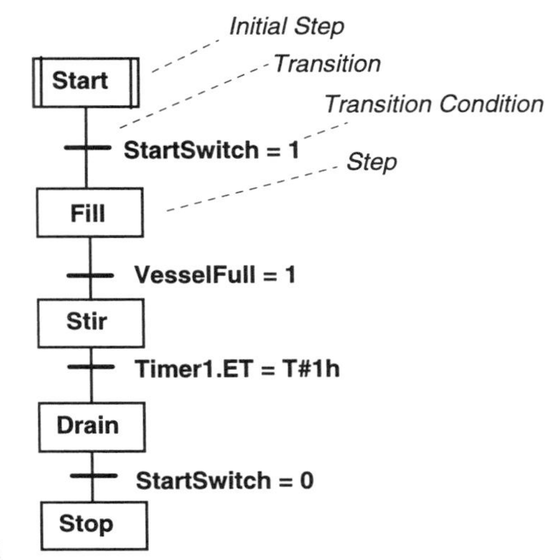

Figure 8.1 Main SFC features

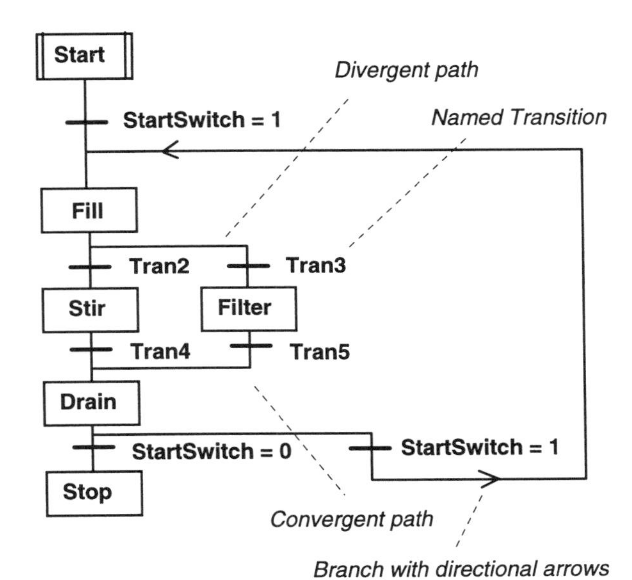

Figure 8.2 Divergent and convergent paths

The example in Figure 8.1 depicts part of a sequence that might be used to describe the main phases of a process controlling a reactor vessel.

The 'Start' step is concerned with holding the vessel in a ready state waiting for the operator to press the start switch. The 'Fill' step becomes active after the operator has initiated the process by setting the StartSwitch to the 'On' state. As the 'Fill' step is activated, the 'Start' step is deactivated in the same instant.

Figure 8.2 depicts the use of divergent and convergent paths in SFCs to provide alternative sequences. In this example, there are two alternative process phases 'Stir' or 'Filter' which could, for example, be selected according to a batch type entered by the operator. Tran4 and Tran5 represent named transitions where the transition condition is defined in another part of the SFC.

These transitions will typically define conditions that will cause the process step to terminate; for example, this could be determined by the process elapse time or some form of process completion parameter such as viscosity or pressure. When either of the conditions for transitions from the 'Stir' or 'Filter' steps becomes true, the 'Drain' step is activated to collect the product. If the StartSwitch remains 'ON', the process is re-started.

Note that the sequence starts with the initial step 'Start'. All SFC sequences must have an initial step. This should be depicted by a rectangular box containing vertical bars. The initial step is the first step to be activated whenever an SFC is started. The initial step remains active until the condition of the following transition becomes true. In the example, the transition condition following the 'Start' step is described using a Structured Text expression.

An SFC sequence will process through all the linked steps. The flow of active steps is referred to as the 'sequence evolution'.

Each step can be associated with one or more actions. Actions which are described in later sections of this chapter define the behaviour that occurs when the step is active, e.g. an action may start a pump, open a valve and so on.

A transition can either be described directly on the Sequential Function Chart using ST, FBD or LD or can be described on an associated diagram using any of the languages ST, FBD, LD or IL, in which case the definition is associated with the name of the transition. For example, in Figure 8.2, the name 'Tran2' is a transition whose definition is given in an associated diagram.

Alternative steps in a sequence can be selected using a divergence construct where more than one transition is associated with leaving a step. In this example, the conditions of both transitions 'Tran2' and 'Tran3' are evaluated when the 'Fill' step is active. Normally, transition conditions are tested from left to right. We will see later that the order in which transitions are tested can be changed. If the condition associated with any transition in a divergent path becomes true, the current step is deactivated and the step that follows the transition with the true condition is activated.

A divergent sequence can rejoin another sequence using a convergent path. Any number of divergent and convergent paths can be used in a Sequential Function Chart enabling complex sequential behaviour to be depicted.

Branches can be used with directional arrows to indicate where transition paths lead back to earlier steps.

Semigraphic representation

The IEC 1131-3 standard defines a semigraphic representation of SFC symbols using textual characters. This is defined for programming systems that are unable to provide full graphics.

Figure 8.3 depicts part of Figure 8.2 using the IEC semigraphic representation:

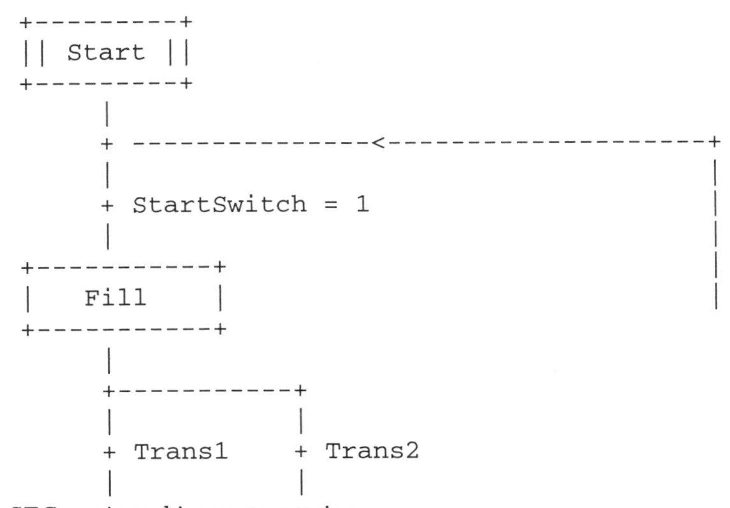

Figure 8.3 SFC semigraphic representation

All the Sequential Function Charts depicted in the rest of this book will use the full graphic form as this is the form most commonly provided by the latest implementations of IEC 1131-3.

Note: The standard does not define how symbols should appear using full graphics. For example, one system may use shaded boxes for steps, another coloured boxes. However, the layout and topology of an SFC should be the same on all IEC compliant systems.

8.3 Simultaneous sequences

We have seen how it is possible to use divergent paths to enter alternative sequences. Using a divergent path, only one step in one of the alternative sequences can be active at any time. However, using a construct, which the IEC standard defines as a simultaneous sequence divergence, it is possible to activate parallel sequences.

This is a particularly useful facility and is commonly required in many batch process applications. The main sequence can be used for sequencing the primary process phases, while secondary parallel sequences can be monitoring that the process is running within normal operating constraints. For example, a parallel sequence could check that plant temperatures and pressures are within operating limits, and, if they are not, initiate a shut-down sequence.

There are situations where the parallel sequences are no longer required, in which case a construct called a simultaneous sequence convergence can be used to bring two or more parallel sequences back into a single sequence.

Figure 8.4 is an example of an SFC which depicts how the behaviour of simultaneous sequences can be depicted using the divergence and convergence constructs; both constructs are depicted by a pair of parallel horizontal lines.

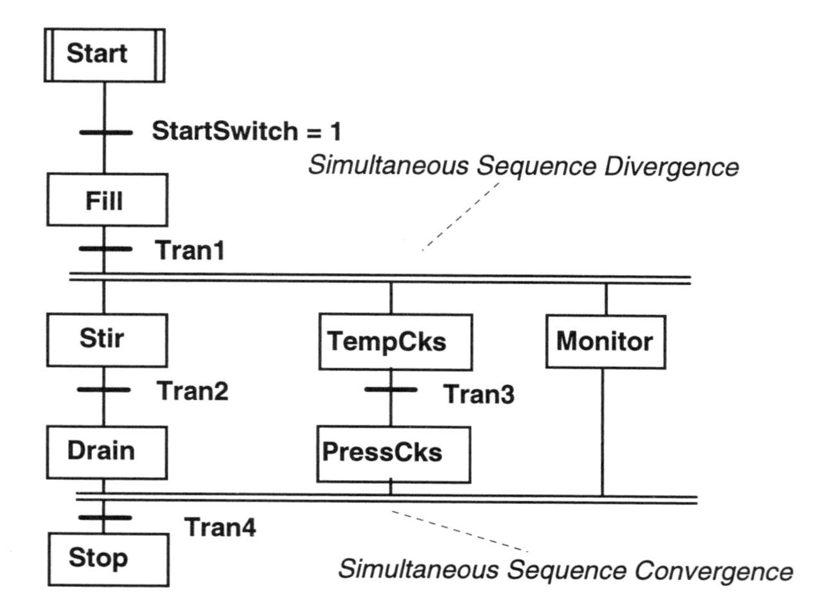

Figure 8.4 Using simultaneous sequences

In Figure 8.4, if step 'Fill' is active while the condition for transition Tran1 becomes true, three steps 'Stir', 'TempCks' and 'Monitor' are simultaneously activated, starting three simultaneous sub-sequences.

Simultaneous sequences continue independently until a convergence is reached. In this case, when all three sub-sequences reach their last step, i.e. when steps 'Drain', 'PressCks' and 'Monitor' are all active and the condition for Tran4 is true, the three steps are deactivated and the following step 'Stop' is activated. The three simultaneous sequences are brought back into one sequence.

Note: The transition closing a simultaneous convergence, e.g. Tran4, is not evaluated until all the last steps of all sub-sequences going to the convergence are active. For example, if steps 'Drain' and 'Monitor' are active but 'PressCks' is not active, Tran4 will not be evaluated.

8.4 Steps

Each step within an SFC should be given a unique name and should only appear once in a Sequential Function Chart. Step and transition names are local to the program organisation unit, e.g. function block or program, in which the SFC exists.

There are two forms of step, both shown in Figure 8.1:

1 **Normal steps:** these are depicted in rectangular boxes with the step name in the centre of the box.

2 **Initial steps:** these are also depicted in identical rectangular boxes but with vertical bars. Only one initial step should be given for any SFC and it defines the step that will be activated when the PLC is cold-started.

The behaviour of any step can be described by associating one or more action blocks with the step. Each action block can be described using any of the IEC languages, i.e. ST, FBD, LD or IL. Action blocks are described in detail in Section 8.6.

Step executing and time variables

Every step is associated with two variables that can be used within the rest of the SFC to synchronise and monitor step activation.

(1) The step active flag (.X) is a boolean variable that is only set true while a particular step is active. This can be tested within other parts of an SFC, for example within other simultaneous sequences, to check that a particular step is active. This variable has the form <StepName>.X. The step box can also provide this variable as a graphical output for driving boolean variables.

For example, in Figure 8.4, the condition for transition Tran3 could contain the ST expression:

```
Drain.X = 1
```

This will result in the transition from step 'TempCks' to 'PressCks' only occurring when the 'Drain' step is activated, i.e. the condition is only true when the 'Drain' step is active.

The following figure depicts how the step active flag (.X) can be connected directly to a boolean variable. Valve1 is only set to 1 when the 'Fill' step is active.

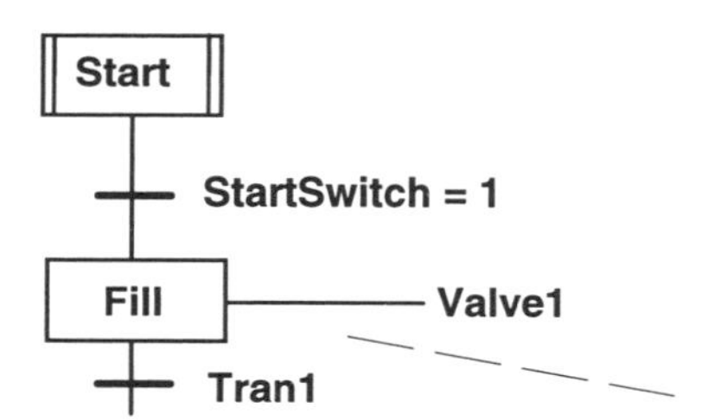

Direct Connection from the step's (.X) active flag.

(2) Each step also has an elapse time (.T) variable of data type TIME. When a step is first activated, the elapse time variable is set to zero. While a step is active, it is updated to indicate how long the step has been active. When the step ceases to be active, the .T variable retains the elapse time, i.e. it records how long a step was active the last time it was activated. The elapse time can be accessed in the form <StepName>.T, e.g. Fill.T.

The elapse time can be very useful for timing how long a step is active. Consider the following example:

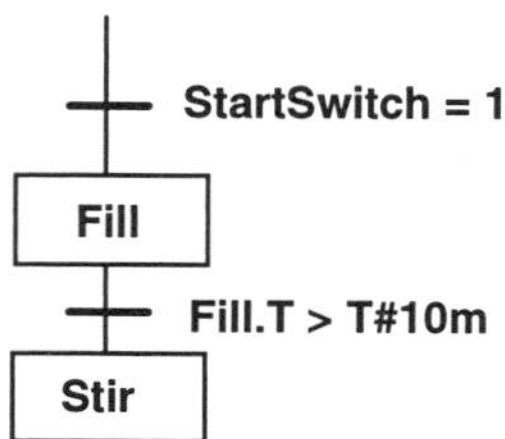

The transition from the 'Fill' step contains an ST expression for the condition 'Fill.T is greater than 10 minutes'. This has the effect of ensuring that the 'Fill' step remains active for exactly 10 minutes. At that point the transition condition is true, the 'Fill' step is deactivated and the next step 'Stir' is activated.

The elapse time can also be used for diagnosing sequencing faults, e.g. to check that all steps are active for their expected times.

8.5 Transitions

The standard provides a wide range of different constructs for describing transition conditions using text or graphics. We have already seen that a transition can be described using Structured Text, e.g. the condition 'StartSwitch = 1' as shown in Figure 8.2.

> Note: Every transition must have a condition. A transition that always occurs can be depicted using a boolean literal 'TRUE'.

The different ways of expressing transition conditions are listed as follows:

Structured Text

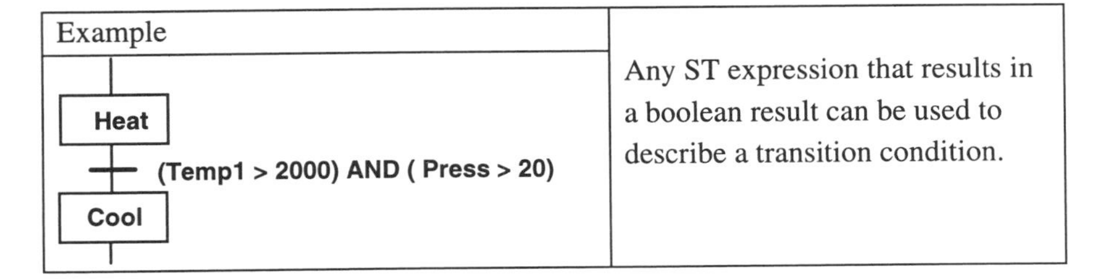

Ladder Diagram

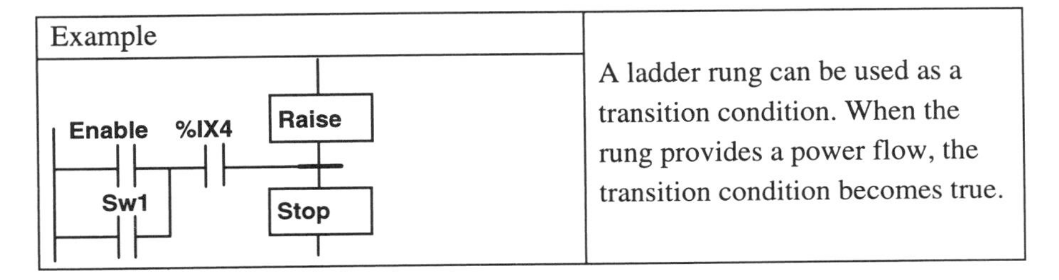

Function Block Diagram

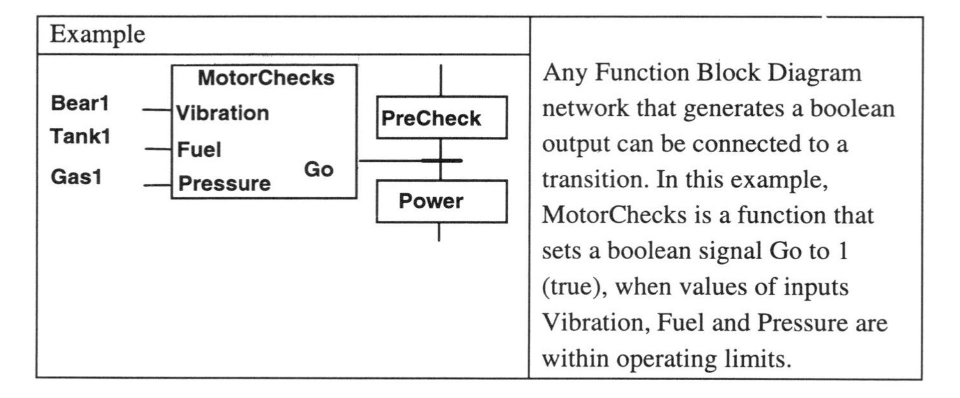

Any Function Block Diagram network that generates a boolean output can be connected to a transition. In this example, MotorChecks is a function that sets a boolean signal Go to 1 (true), when values of inputs Vibration, Fuel and Pressure are within operating limits.

Transition connector

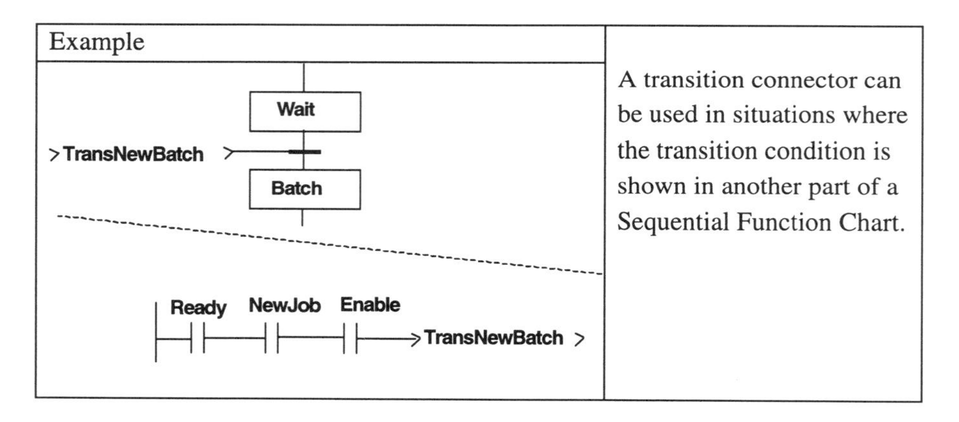

A transition connector can be used in situations where the transition condition is shown in another part of a Sequential Function Chart.

Named transition

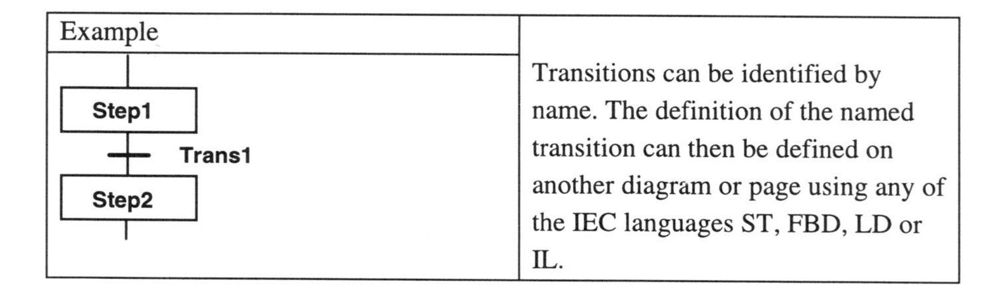

Transitions can be identified by name. The definition of the named transition can then be defined on another diagram or page using any of the IEC languages ST, FBD, LD or IL.

Transition definition using ST

Example	

`TRANSITION Trans1 :`
`    := AB1 AND CX3 OR CX5`
`       AND (TX3 >= 100.2);`
`END_TRANSITION`
	A named transition condition can be described using a Structured Text expression. The expression should only return a boolean value.
	Note that the assignment symbol := is required to indicate that the value is assigned to the transition condition.

Transition definition using FBD

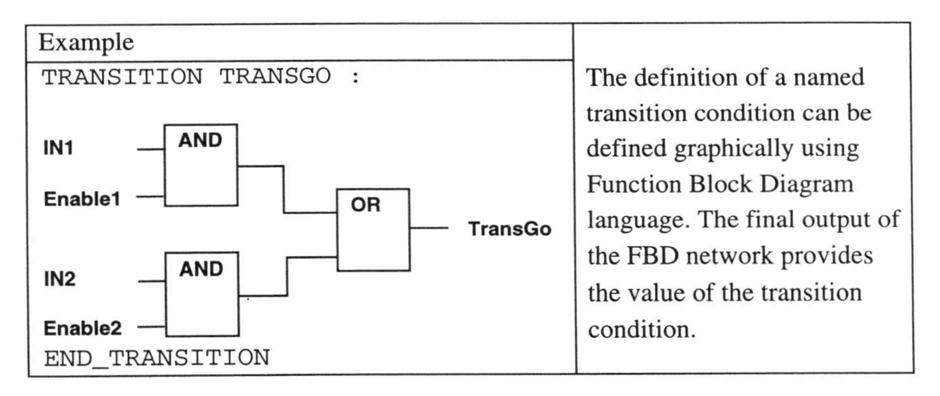

The definition of a named transition condition can be defined graphically using Function Block Diagram language. The final output of the FBD network provides the value of the transition condition.

Transition definition using LD

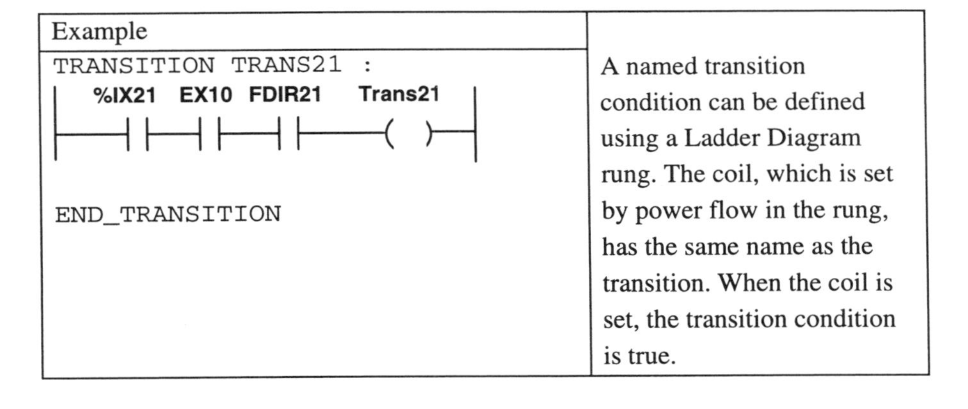

A named transition condition can be defined using a Ladder Diagram rung. The coil, which is set by power flow in the rung, has the same name as the transition. When the coil is set, the transition condition is true.

Transition definition using IL

Example	

`TRANSITION Trans21 :`
`    LD  %IX21`
`    AND EX10`
`    AND FDIR21`
`END_TRANSITION`
	A named transition can also be defined using the Instruction List language. The value of result register after executing the IL instructions defines the value of the transition condition. When the value, that should be boolean data type, is 1 the transition condition is true.

8.6 Actions

Up to now we have only considered steps as being used to describe the states of a plant or machine. In a real system, each step is required to perform certain control actions that are applicable for that particular state.

In SFC methodology, every step can be associated with one or more actions. An action contains a description of some behaviour that should occur as a result of the step being activated. An action is depicted as a rectangular box that is attached to a step. The description of an action can be given using any of the IEC languages, i.e. ST, FBD, LD or IL.

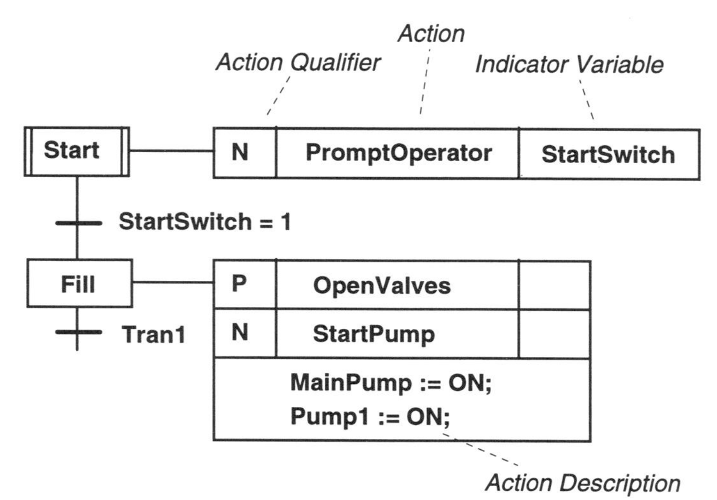

Figure 8.5 Main features of SFC actions

Figure 8.5 depicts the main features of SFC actions. Every action has a name which should be unique within the current program organisation unit, e.g. unique within the current function block that contains the SFC.

An action can either be described on another diagram or page or, if its behaviour is fairly trivial, it can be described as part of the SFC, as shown for the action 'StartPump' in Figure 8.5. Any of the IEC languages can be used to describe the behaviour of an action. There is no restriction within the standard on the level of complexity of networks or number of language statements used to define an action.

Each action has a qualifier that determines when the action is executed. In this example, the 'N' qualifier indicates that the action is executed while the associated step is active. A range of qualifiers is defined in the standard and is described in the next section.

An action box may optionally have an indicator variable. This is only used for annotation purposes. The indicator variable names a key variable that is changed within the action and indicates that the action has completed its execution. In Figure 8.5, action 'PromptOperator' signals the operator that the process can be started. As a result, the operator sets a switch called StartSwitch to 1. StartSwitch is therefore shown as a key variable of the 'PromptOperator' action and indicates that it has finished.

Any step may have any number of actions or none. A step with no actions simply waits for its transition conditions to become true; it performs a 'wait' function.

An action may also be associated with more than one step. For example, the action 'OpenValves' shown in Figure 8.5 could also be used in other process steps, such as in a 'CLEANING' step concerned with setting valve positions for cleaning the reactor, i.e. wherever the same valve operations are required the same action can be called.

The behaviour of some actions may be too complex to draw directly on the SFC diagram, in which case the behaviour can be defined on another diagram or page using any of the IEC languages including SFC.

The following examples show how the various IEC languages can be used to describe actions.

Action defined using Ladder Diagram

Example	
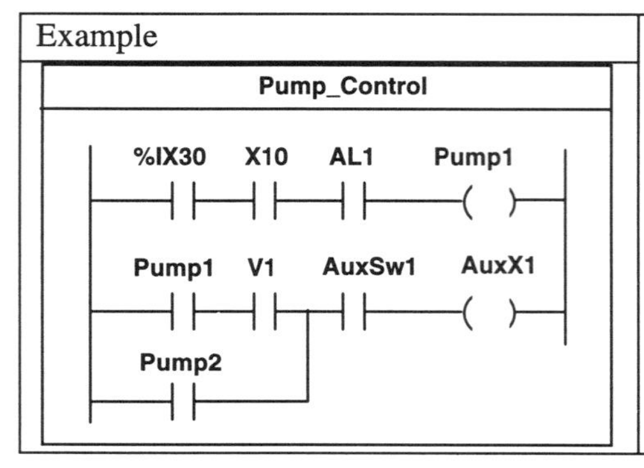	The behaviour of an action can be described graphically using the Ladder Diagram language. The rungs are enclosed in a rectangular box, with the name of the action placed at the top.

Action defined using Function Block Diagram

Example	
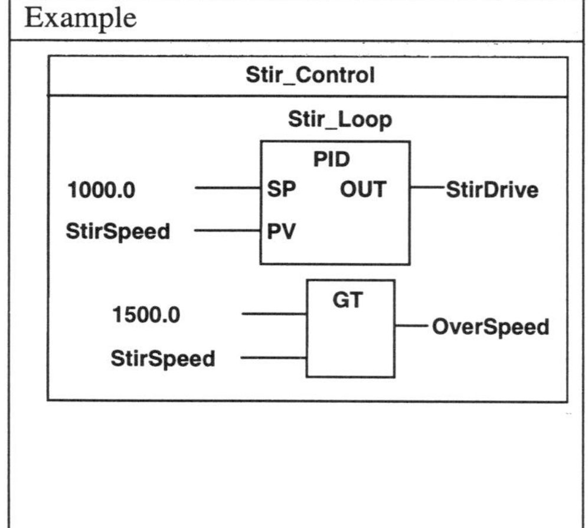	The behaviour of an action can be described graphically using the Function Block Diagram language. The FBD network is enclosed in a rectangular box, with the name of the action placed at the top. Note: If an action is configured to execute continually it can be used to contain feedback control loops. For example, an action could be used to bring a set of PID control loops into operation during a particular process phase.

Action defined using Sequential Function Chart

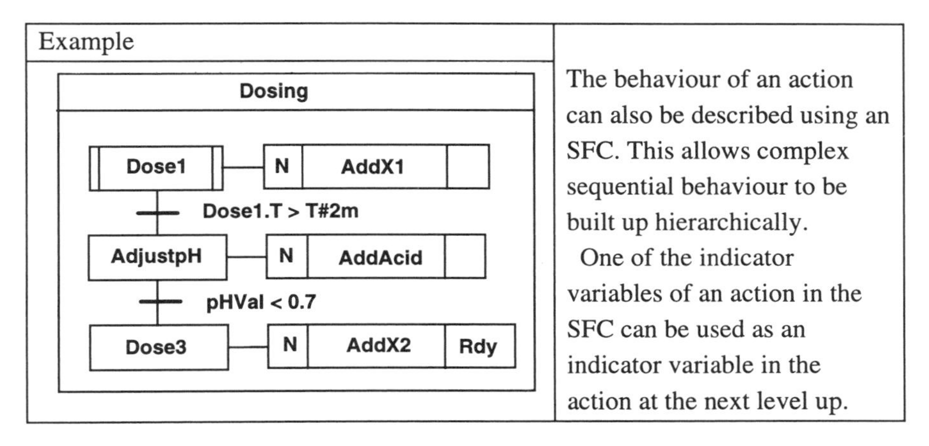

Note: For an action to be able to execute its own SFC, it should be configured to execute continually for a period of sufficient duration to allow its SFC to complete.

Action defined using Structured Text

Example

```
ACTION OpenValves :
   IF BatchType = "AB0_100" THEN
      ValveAB := OPEN;
   ELSE
      ValveAC := OPEN;
   END_IF;
   DrainValve := CLOSE;
   Vent := OPEN;
END_ACTION
```

The behaviour of an action can be described textually using a series of statements in Structured Text.

A variable changed by an ST statement can be used as the action indicator variable.

Action defined using Instruction List

Example	

`ACTION AddX2 :`
`    LD     1`
`    ST     TankX2.Enable`
`    LD     100`
`    ST     TankX2.ShotLevel`
`    CAL    TankX2`
`    S      X2_Added`
`END_ACTION`
	Actions can also be described textually using Instruction List.
	A variable changed by one of the IL instructions can be used as the action indicator variable.

Use of actions in graphical languages

The use of actions is not restricted to SFC. Actions can also be called within FBD and LD languages as shown in Figure 8.6

In the FBD language, the action is activated when its boolean input is true. In LD, the action is activated when there is a power flow into the action block. In both languages, the action indicator variable can be used to signal when the action is complete.

Although the standard allows actions to be used in the graphical languages in this way, such use is not recommended. Functions and function blocks with defined inputs and outputs provide a better structure for partitioning program behaviour.

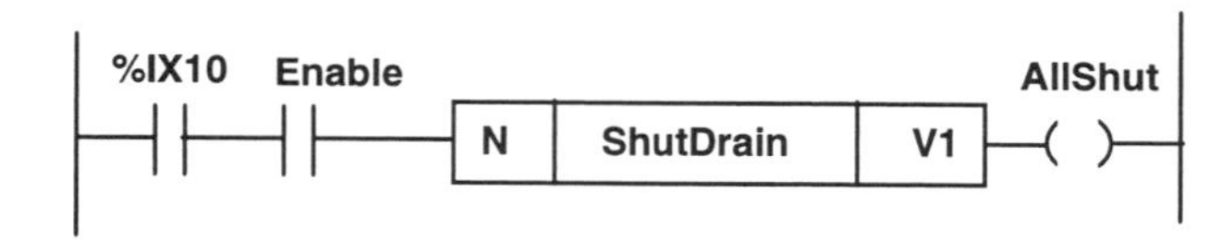

Using Action Blocks in Ladder Diagram

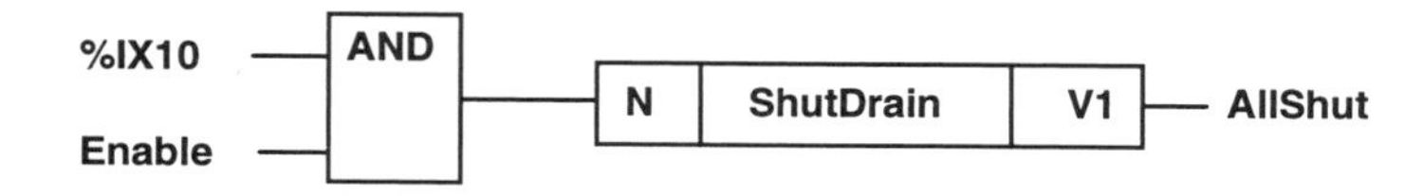

Using Action Blocks in Function Block Diagram

Figure 8.6 Using actions in graphical languages

8.7 Action qualifiers

So far we have considered actions which are configured to execute continually while their associated steps are active. Such actions all have the 'N' qualifier. Remember that the qualifier is always given in the first field of the action block as shown in Figure 8.5.

The standard defines a range of qualifiers which define precisely when a particular action executes in relation to its associated step.

For example, the 'D' qualifier (time delayed) allows an action to start execution a defined time after its associated step is activated.

An action can also be set active, using an 'S' qualifier, in one step and continue to execute until reset, using the 'R' qualifier, in another step. This may occur many steps, or some considerable time, later. Such actions that persist for more than one step are said to be 'stored'. The full range of qualifiers defined in the standard is given in Table 8.1.

Table 8.32 SFC action qualifiers

Qualifier	Description
None	Non-stored, default, same as 'N'.
N	Non-stored, executes while associated step is active.
R	Resets a stored action.
S	Sets an action active, i.e. stored.
L Note 1	Time limited action, terminates after a given period.
D Note 1	Time delayed action, starts after a given period.
P	A pulse action that only executes once when a step is activated, and once when the step is deactivated.
P1 Note 2	A pulse action that only executes once when a step is activated.
P0 Note 2	A pulse action that only executes once when a step is deactivated.
SD Note 1	Stored and time delayed. The action is set active after a given period, even if the associated step is deactivated before the delay period.
DS Note 1	Action is time delayed and stored. If the associated step is deactivated before the delay period, the action is not stored.
SL Note 1	Stored and time limited. The action is started and executes for a given period.

Note 1: These qualifiers all require a time period.
Note 2: These action qualifiers are proposed in the 1998 IEC 1131-3 amendments for the second edition.

Care should be taken selecting action qualifiers to ensure that actions only execute when required. The behaviour of all stored actions, that continue to execute long after their associated steps have been deactivated, requires careful scrutiny.

The behaviour of the different qualifier types is best understood from the following timing diagrams:

'N' non stored action qualifier

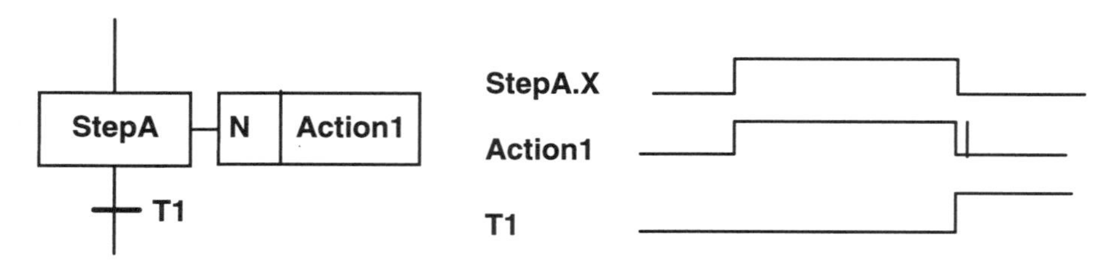

The action 'Action1' executes continually while 'StepA' is active, i.e. while the step's .X executing flag is 1. It then executes one last time after the step is deactivated.

'S' set and 'R' reset action qualifiers

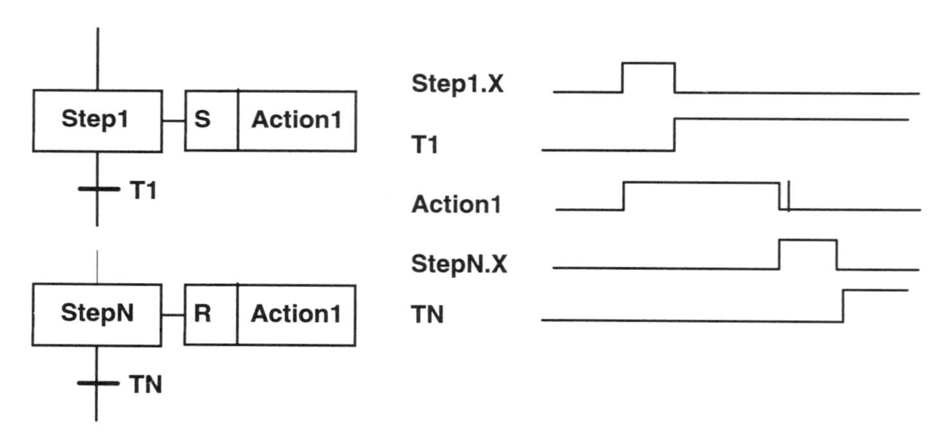

The S qualifier causes the action 'Action1' to begin execution immediately the step becomes active; the action is stored. The action continues to execute until an action block with an R qualifier that references 'Action1' is reached. This resets the stored action and it ceases one execution after the step 'StepN' is activated. If a stored action is never reset, it will continue to execute indefinitely.

'L' time limited action qualifier

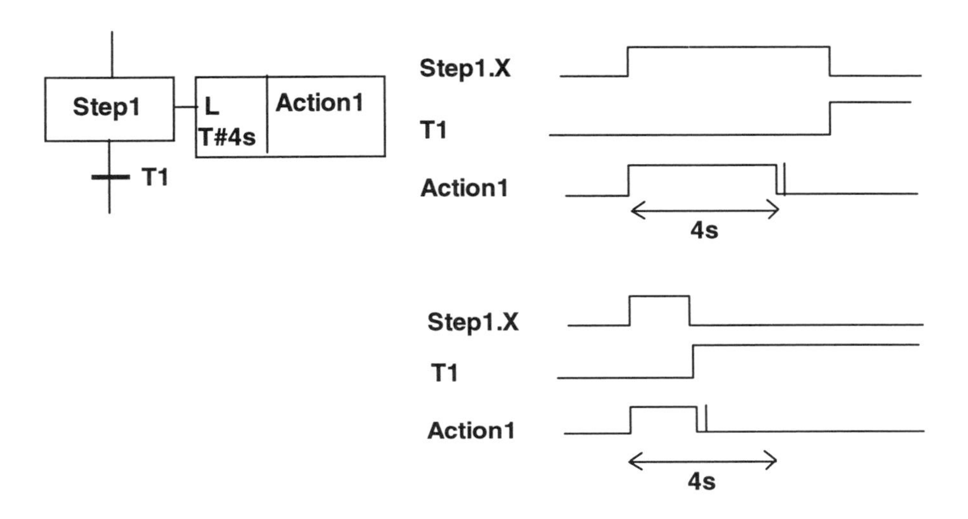

The action 'Action1' starts to execute when 'Step1' becomes active. It continues to execute until a given period has elapsed. If the step is deactivated before the action's period has elapsed, the action ceases execution. In both cases, the action executes one last time, i.e. either after the period has elapsed or after the step has been deactivated.

'D' time delayed action qualifier

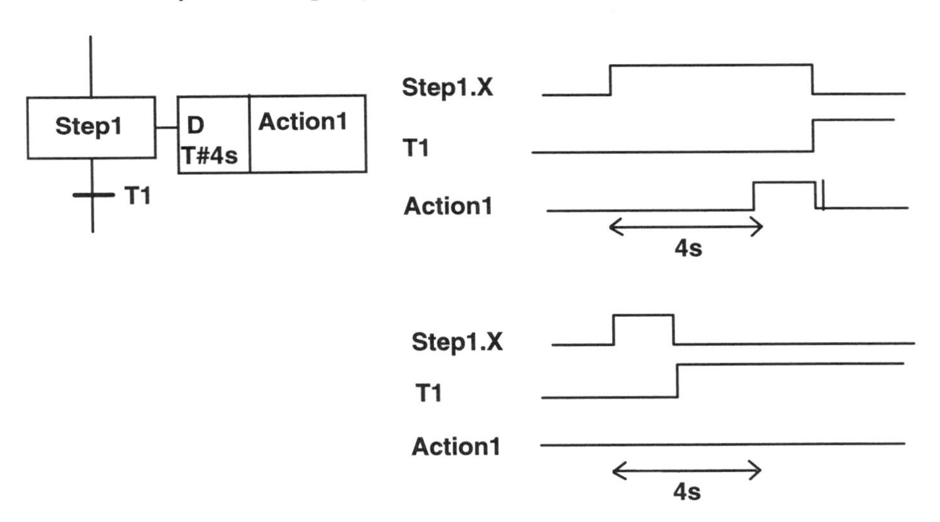

When the given time period has elapsed, timed from when step 'Step1' becomes active, the action starts to execute. It continues to execute while 'Step1' is active. It then executes one last time after the step is deactivated. If the step is not active for longer than the delay period, the action does not execute.

'P' pulse action qualifier

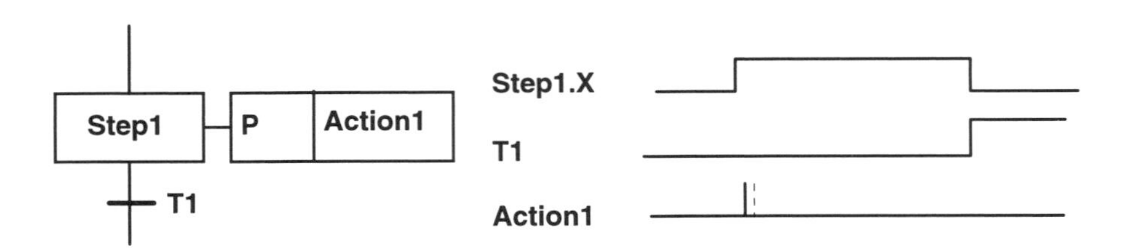

With the P qualifier, the action executes once when 'Step1' is activated.

Note: If the standard is implemented exactly as written, the pulse action 'P' also executes a second time from the rule that action statements or networks shall be executed one final time after the falling edge of the action qualifier. In some implementations of IEC 1131-3 the pulse action is not executed a second time on the falling edge of the action qualifier. This may lead to a difference in program behaviour when ported to different systems.

'P1' and 'P0' pulse action qualifiers

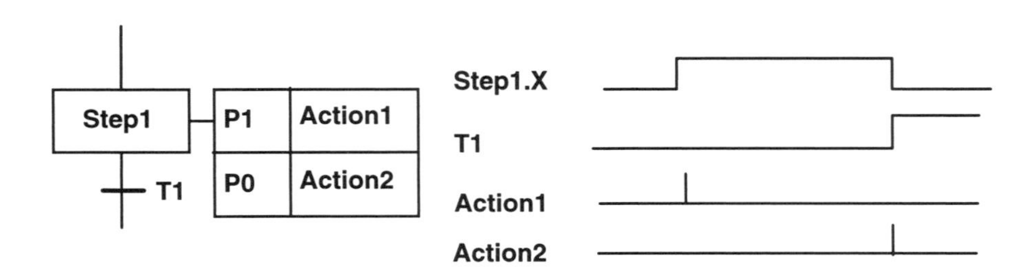

The IEC 1131-3 amendment for the second edition defines two additional action qualifiers P0 and P1. With the P1 qualifier, the action executes once when 'Step1' is activated; the P0 qualifier executes the action once when the step is de-activated. These should be used in preference to the 'P' qualifier.

'SD' stored and time delayed action qualifier

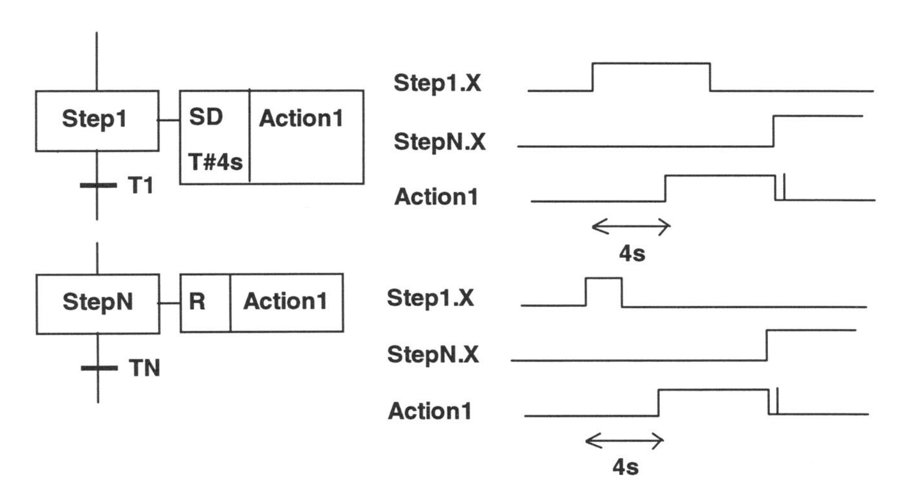

On 'Step1' being activated, the action is stored but does not start to execute until the time period has elapsed. The action continues to execute until it is referenced in a later step with an R reset qualifier. Even if 'Step1' is deactivated before the time period has elapsed, the action will still start to execute after the time period.

However, the stored action will not execute, if the same action is called with R qualifier, before the time period has elapsed.

'DS' time delayed and stored action qualifier

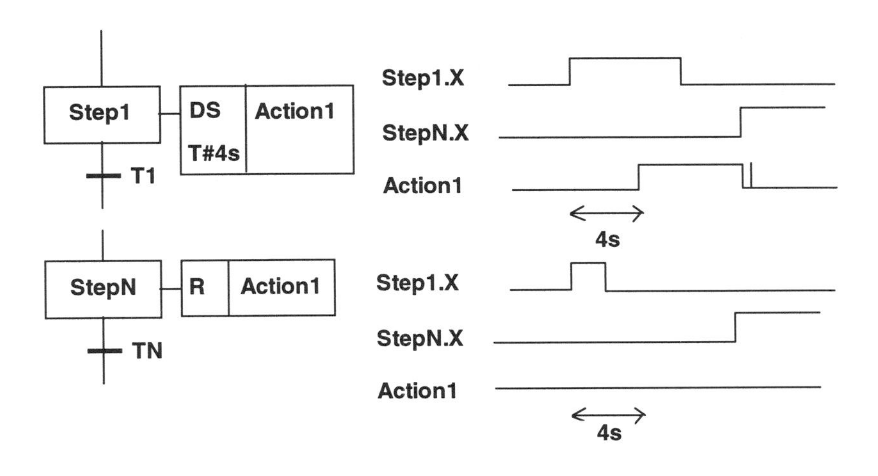

On 'Step1' being activated, the action is delayed. After the time period has elapsed, the action is stored and starts to execute. The action continues to execute until it is referenced in a later step with an R reset qualifier. If 'Step1' is deactivated before the time period has elapsed, the action will not be stored and it does not execute.

'SL' stored and time limited action qualifier

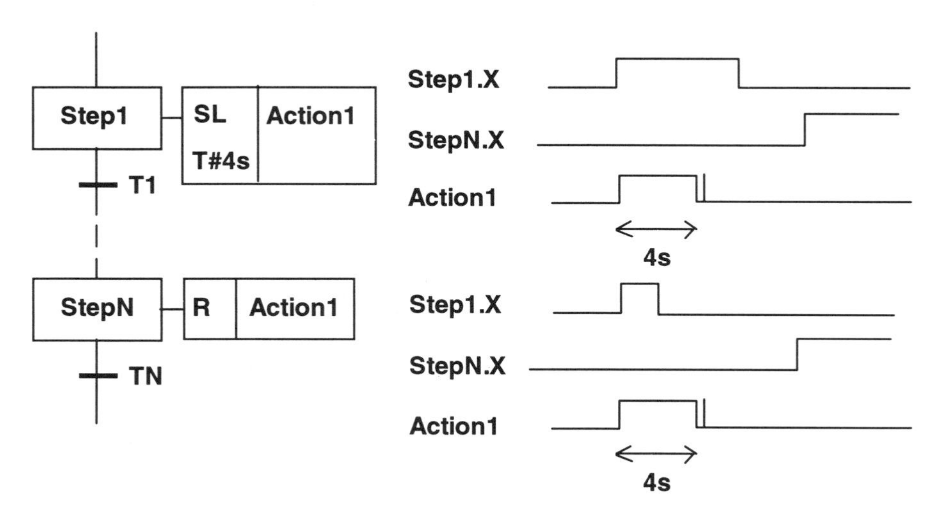

When step 'Step1' becomes active, the action 'Action1' is stored and begins to execute. It continues to execute until the time period has elapsed. Even if 'Step1' is deactivated before the period has elapsed, the action still executes for the complete time period.

However, the stored action will cease execution, if the same action is called with R qualifier, before the time period has elapsed. In all cases, the action will execute one last time either after the period has elapsed or after the stored action has been reset.

The action_control function block

The standard defines a conceptual action control function block to describe the behaviour of all action qualifiers. Each action is assumed to be controlled by one instance of this function block. The action control function block does not actually exist! It is defined in the standard purely as a means of describing the action execution behaviour. The output of the action control function block is called 'Q' and when true results in an action being executed.

The standard states that while Q is true, the action associated with the control function block is executed. It also states that all action 'statements or networks shall be executed one final time after the falling edge of Q'. As a result, actions execute once more after the action's associated qualifier is cleared.

This feature has led to some misunderstanding of how actions execute. The extra execution, that occurs after the associated action has deactivated, can lead to some subtle and unexpected side-effects.

For example, consider an action with a 'P' pulse qualifier, defined in Structured Text as:

```
ACTION COUNTING :
      COUNT := COUNT + 1;
END_ACTION
```

Assume this is an action that is associated with a step and is called A1. Every time A1 is active, COUNT is incremented by one. However, every time A1 is deactivated, COUNT is incremented again due to the rule that action statements 'execute one final time on the falling edge of Q'.

The standard has provided this feature so that an action can reset certain variables after an action block has been deactivated. However, the double execution of the 'P' action is an oversight. An amendment to the 1993 version of the IEC 1131-3 standard provides access to the action qualifier (.Q). It is then possible to test when an action is executing for the last time because Q will be false.

It is possible by using certain function blocks that are triggered from the rising edge of the action qualifier (.Q) to ensure that selected operations only occur once within action blocks.

For example it might be necessary to count the number of times a motor is switched on. This can be achieved in a single action block using the 'P' pulse qualifier. Using an Up-counter function block that counts on the rising edge of the 'CountUp' action qualifier, ensures that the count only increases once.

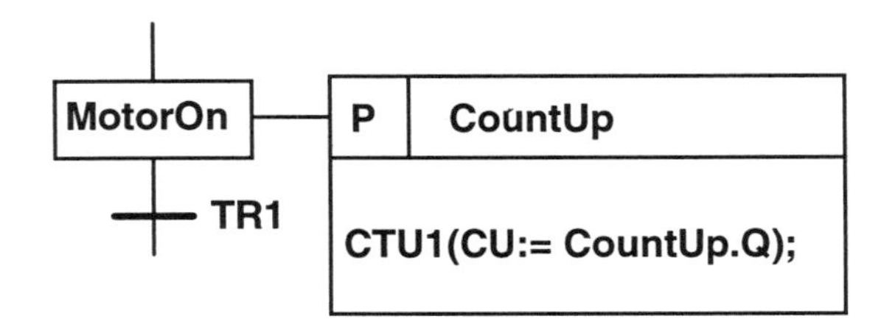

Although the 'CountUp' action is executed twice (once as the P action is activated, once when it is deactivated), the 'Up-counter' only counts once as it detects the rising edge on input CU.

Note: The P0 and P1 action qualifiers defined in the IEC 1131-3 amendment now solve this problem because the action is only executed once.

8.8 SFC execution

SFC networks can be used to describe both the behaviour of program organisation units (POU), i.e. programs and function blocks, and the behaviour of actions (see Section 8.6). In both cases, the execution of the associated SFC network follows the same rules.

Every program organisation unit is associated with a task. The task is responsible for periodically executing the language elements within the POU. Generally, POUs are executed at a regular rate, for example every 500 ms. An active action executes at the same rate as the SFC that contains it.

An SFC within a POU will be evaluated every time the POU executes. Similarly, an SFC within an action will be evaluated every time the action executes.

The SFC evaluation rules are:

---

1. The initial step of an SFC is always activated when the POU executes after system initialisation. Any actions associated with the initial step are executed.
2. At the start of each evaluation, the current set of active steps is determined. All transitions that are associated with active steps are evaluated.
3. Actions which nominally ceased execution in the previous SFC evaluation are executed one last time.
4. All actions which are active are executed.
5. Any active steps that precede transition conditions that are true are deactivated and their succeeding steps are activated.
6. Any actions, having an execution enabling condition that ceases, are marked as inactive, for example, an action with a time limited 'L' qualifier when the time period has elapsed. Such actions are executed one last time in the next SFC evaluation.

8.9 SFC rules of evolution

The IEC 1131-3 standard defines a number of constraints on the way SFC networks are designed and interpreted to ensure that the behaviour is unambiguous.

---

1. Two steps can never be directly linked; they should always be separated by one transition.
2. Two transitions can never be directly linked; they should always be separated by a step.
3. If a transition from one step leads to two or more steps, then such steps initiate simultaneous sequences. Simultaneous sequences continue independently.
4. When designing an SFC, the time to clear a transition, deactivate preceding steps and activate succeeding steps may be regarded as occurring instantaneously.
5. Where more than one transition condition is true at the same time, all the deactivations and activations of associated steps appear to occur in zero time. The designer should not consider the effects due to timing differences between transitions that occur simultaneously.
6. A transition condition of a step is not evaluated until all the behaviour resulting from the active step has propagated through the program organisation unit. For example, if a step is activated that has a transition condition that is always true, all the applicable actions are executed once before the step is deactivated.

Note: In a PLC the time to process SFC steps and transitions may become significant, especially if the SFC is running in a very fast task.

Evolution of divergent paths

Where there are two or more transitions from one step, the sequence can diverge to one of a selected number of steps. Even if more than one transition condition is true simultaneously, only one path is selected. The selected path is determined by the precedence of the transition.

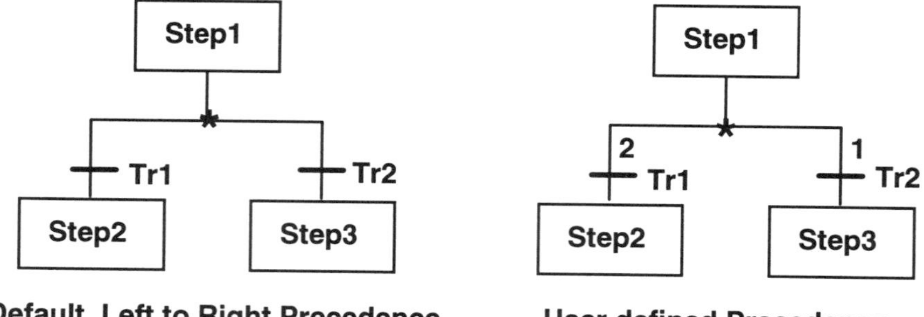

Figure 8.7 Precedence of divergent paths

The standard allows the precedence of transition conditions to be modified. Consider the transitions shown in the left SFC of Figure 8.7. Left to right evaluation is the normal default. If transitions Tr1 and Tr2 are both true simultaneously, the transition is always made to 'Step2', i.e. the leftmost step. The default transition priority may be shown in SFC networks by an asterisk placed at the centre of the divergent branches. If an asterisk is not present, the default left to right evaluation is always assumed.

It is possible to override the default precedence by numbering the transitions. The transition with the lowest number then has the highest priority. Consider the SFC shown on the right of Figure 8.7; if both Tr1 and Tr2 are true simultaneously, then the transition is made to 'Step3'. User defined precedence is denoted by numbered transitions and placing an asterisk at the centre of the branches.

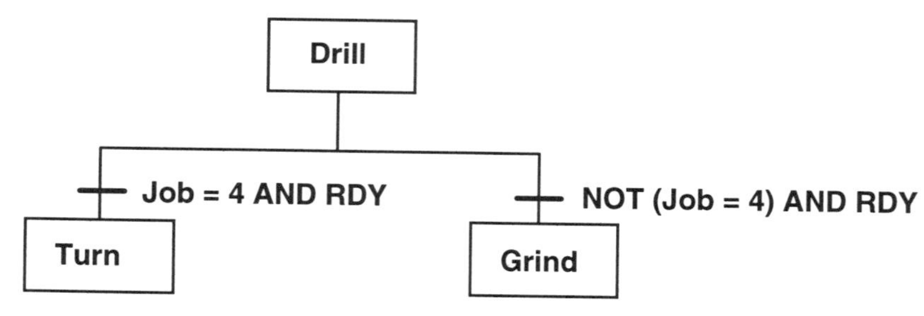

Figure 8.8 Mutually exclusive divergent paths

To avoid ambiguous situations that may occur when transitions become true simultaneously, it is good practice to ensure that transition conditions in divergent paths are always mutually exclusive, as shown in Figure 8.8. This ensures that the flow through an SFC network is easier to follow and complies with the requirements of the original function chart standard IEC 848.

Sequence skip

A divergent path may also be used to skip a section of a sequence. In this situation the divergent path contains no steps, as shown in the SFC on the left side of Figure 8.9.

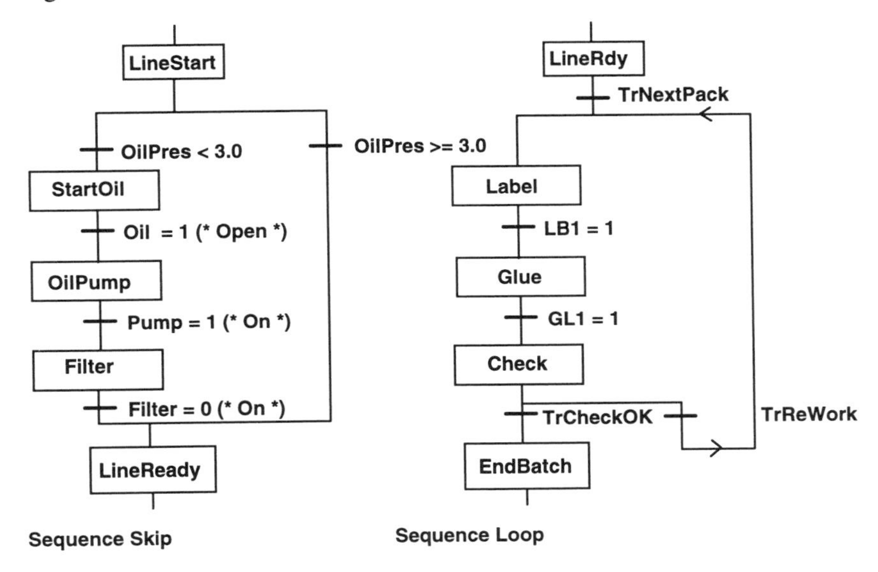

Figure 8.9 Special forms of divergent paths

Sequence loop

A divergent path can also be used to repeat a section of sequence by branching back to an earlier step. If required to add clarity, arrows may be used to indicate the direction of flow in branches, as shown on the right side of Figure 8.9.

Simultaneous convergence

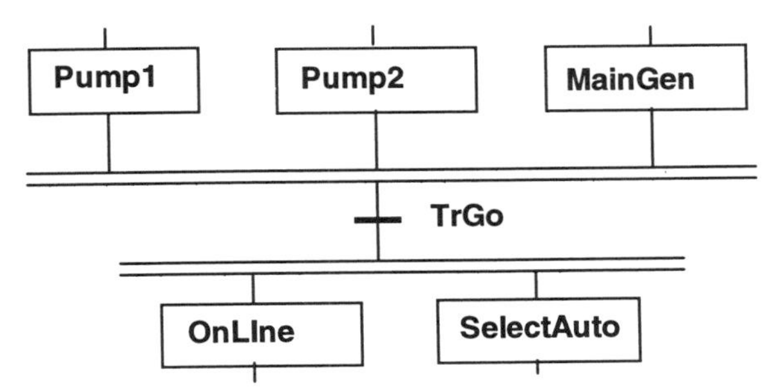

Figure 8.10 Convergent simultaneous sequences

A number of simultaneous sequences can converge to a single transition, and then diverge to a different number of simultaneous sequences. This construct can be regarded as a 'rendezvous'. All simultaneous sequences going into such a transition must be complete before the transition condition is evaluated. In Figure 8.10, steps 'Pump1', 'Pump2' and 'MainGen' must all be active for the transition to be tested.

At the point when all preceding steps are active and the transition is cleared, all preceding steps are then deactivated and the succeeding steps activated. This all happens in one instant.

8.10 Safe and unsafe design

Certain arrangements of branches in SFC networks can lead to unsafe designs. Situations where a divergent path branches out of a simultaneous sequence can lead to sequences that never complete or that behave unpredictably.

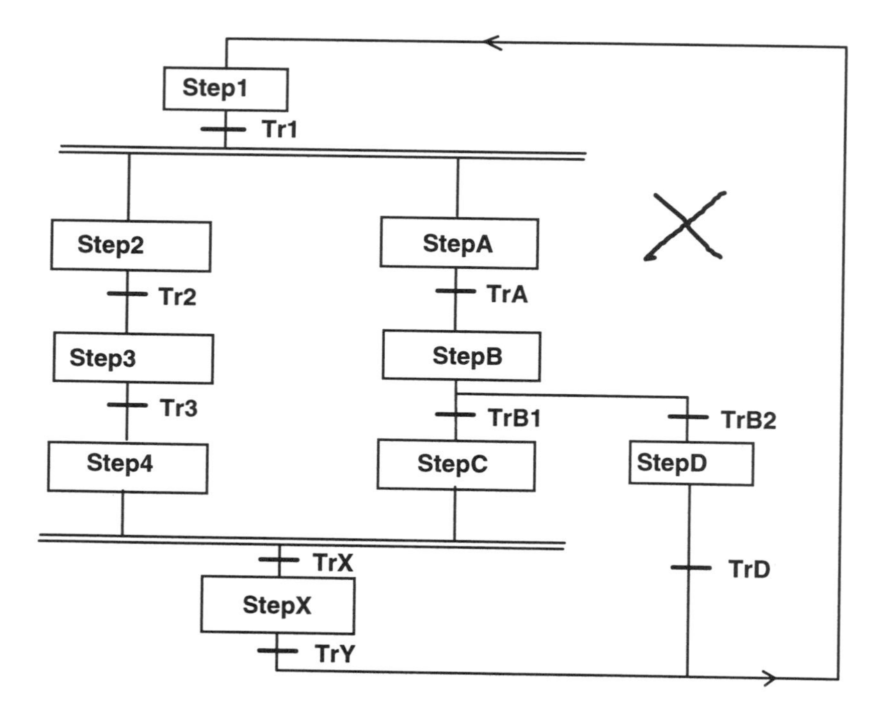

Figure 8.11 Unsafe SFC

Figure 8.11 depicts an unsafe SFC network. If the divergent path to 'StepD' is taken, the sequence execution returns to an earlier step 'Step1'. As a consequence, one of the steps within the sub-sequence 'Step2' to 'Step4' will remain active. As execution of the SFC continues, it is possible for more than one step to become active in the same sub-sequence, resulting in unpredictable behaviour.

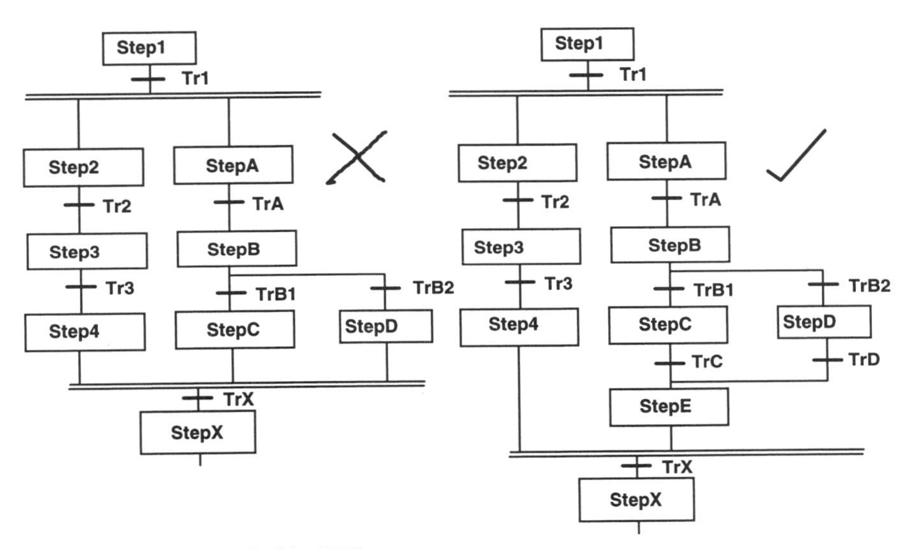

Figure 8.12 Unreachable SFC

Figure 8.12 depicts another form of unsafe SFC. In the SFC on the left, 'StepX' is unreachable because transition TrX is never evaluated. This is because steps 'StepC' and 'StepD' are mutually exclusive; it is never possible for all the steps going into the simultaneous sequence convergence to be active. By adding a further step as shown on the right, transition TrX can be evaluated and step 'StepX' reached. Although step 'StepE' may contain no actions, it is required to indicate that the sequence initiated by step 'StepA' has completed.

Generally the same number of sequences that are initiated by simultaneous divergence should terminate at a simultaneous convergence.

Some programming systems are able to analyse SFC networks and check for unsafe constructs. However, care should always be taken, particularly when using simultaneous sequences, to avoid unpredictable behaviour.

Detecting unsafe SFCs

It is possible for an SFC compiler to detect some forms of unsafe SFC construct by repeatedly applying the following simple algorithm. A complete chart should resolve down to a single step. If it does not, the SFC has an unsafe design.

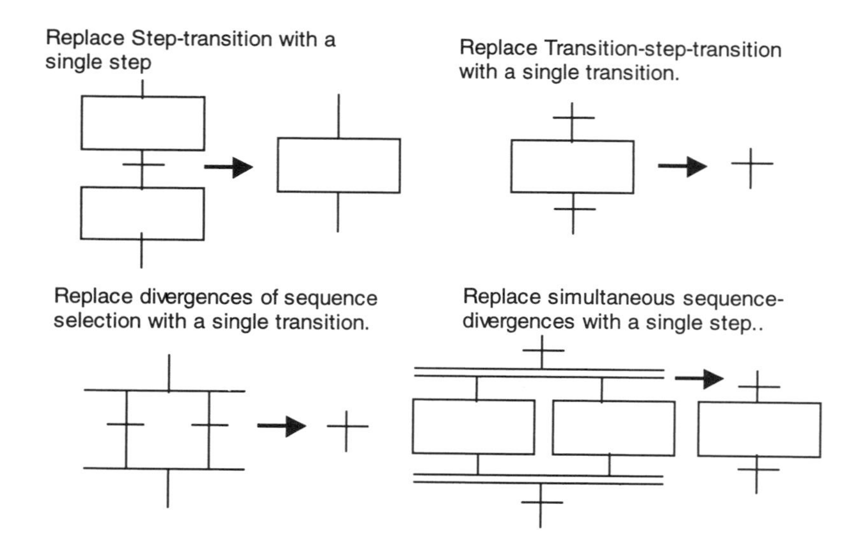

8.11 Top down design

One of the main uses of Sequential Function Charts is as an aid to developing the top down design of a complex process or operation. SFC is primarily used to describe the main changes in state that occur over time. Whenever designing the top level SFC for any control system, start by identifying the main states.

For example, consider the design of a furnace for the heat treatment of steel ingots. This might have the following main states:

'IDLE'	The furnace is not in use and cold
'HEATING'	The furnace is being heated to the working temperature
'EMPTY'	The furnace does not contain any ingots
'LOADING'	The furnace door is open; the ingot is being moved into position
'READY'	The furnace contains an ingot; the furnace door is shut
'TREATING'	The furnace is loaded and at the working temperature
'FINISHED'	The heat treatment process is complete and ingots can be unloaded
'COOLING'	The furnace is cooling from its working temperature

Although these are the main states, there are typically many minor states that exist within each of the main states. For example, during the 'TREATING' state it may be necessary to cycle through a number of different temperatures represented by states such as 'RAMP TEMPERATURE', 'HOLD TEMPERATURE', etc.

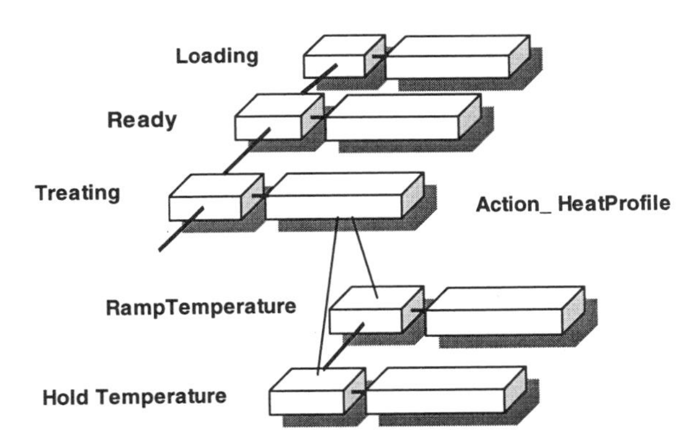

Each process or plant state can normally be represented as a step in an SFC. It is good design practice to represent minor states that exist within the main process states, as steps in an SFC at a lower level in the design hierarchy. This can be achieved by creating an SFC within an action block, encapsulating the SFC in a function block or using a macro step (see Section 8.12). For example, the 'TREATING' step might have an action called 'HEATPROFILE' which calls an SFC that generates a profile of temperatures.

Why the process changes from one state to another should also be considered. The change from 'IDLE' to 'HEATING' may occur, for example, because a new ingot has been delivered to the furnace and the operator has initiated a treatment cycle. The reasons for the changes of state can be translated into transitions. Each transition generally occurs because of an event such as the operator activating a switch, the furnace reaching a working temperature, or a period of time elapsing.

The SFC allows the control problem to be broken down hierarchically into steps and smaller sequences that can then be analysed in detail. It is then possible to focus on the behaviour in each particular state.

8.12 Macro steps

Although not defined in the standard, many implementations of IEC 1131-3 also support the concept of macro steps to help structure large SFCs. A macro step can be used to hide a complex section of a Sequential Function Chart at a lower level. This allows a large SFC to be broken down into a number of layers without changing the way the complete SFC executes.

Consider Figure 8.13. 'Conveyor' is an example of a macro step that contains a complex section of SFC. In effect, when the SFC executes, the macro step is replaced by the 'Conveyor' macro step definition. In this case, the step evolution into the macro is from 'LineStart' directly to 'ConvStart' and then back from 'Ready' to 'Feeds'. It is important to note that although the transition condition following a macro step may become true while the macro step is active, the transition will not be made until the last step in the macro definition has been reached.

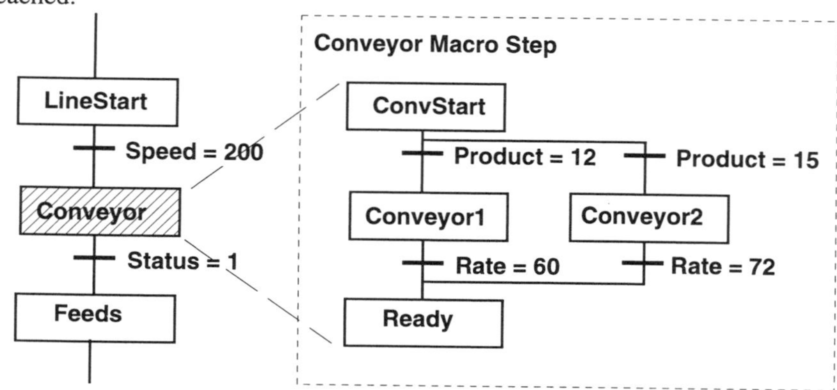

Figure 8.13 Macro step example

Note: Macro steps can be regarded as a means to organise large SFCs. They do not change the way the SFC executes. However some implementations have added additional run-time support for macro steps e.g. diagnostics for showing macro step status.

8.13 Good programming style

To ensure that Sequential Function Charts are always easy to follow and maintain, the following points should always be considered:

- Use meaningful names for steps, transitions and actions wherever possible.
- Try to keep each SFC small and focused on the main behaviour being considered at that level in the design hierarchy; all detailed behaviour should be contained within action blocks and depicted in SFCs at lower levels.
- Try to reduce the interaction between different simultaneous sequences to a minimum.
- Avoid allowing actions of steps within different simultaneous sequences from changing the same variables. For example, avoid steps in different sequences from changing the same function block input parameter.
- Care should be taken where an action contains a sub-sequence that can be stopped in any state by the 'top' level sequence. The 'top' level sequence must always ensure that all incomplete operations arising from suspended sub-sequences are correctly 'tidied-up'. This problem commonly arises when sequences are prematurely halted by some form of plant trip condition.

8.14 Scope of names within SFC networks

All names for steps, transitions, transition connectors and actions should be unique within the program organisation unit (e.g. function block) that contains the SFC network.

However SFC networks within function blocks, such as those called within action blocks, are limited to the scope of their associated program organisation unit. In other words, there may be many steps called 'Start' provided that they all exist in different function blocks or programs.

Note: SFC elements within an action block are considered to be within the same scope as the outer SFC.

8.15 SFC textual form

Sequential Function Charts can also be described textually. The standard provides a language syntax so that a complete SFC can be described in textual statements. The textual form can be used for storing SFCs as text. One of the main uses could be as an intermediate form for porting SFC designs between different PLC systems.

If all actions and transitions are also described using a textual language, i.e. Structured Text or Instruction List, the complete SFC design can be described in a set of text files that can be imported into other systems. In the future, software analysis tools may be developed to check the SFC structure formally from the textual SFC definition.

The following set of text statements describe the fragment of SFC shown in Figure 8.5.

```
INITIAL_STEP Start:
 Prompt_Operator(N,StartSwitch);
END_STEP

TRANSITION FROM Start TO Fill :
  StartSwitch = 1;
END_TRANSITION

STEP FILL :
 OpenValves(P);
 StartPump(N);
END_STEP

ACTION StartPump :
  MainPump := ON;
  Pump1 := ON;
END_ACTION
```

The step body between the keywords STEP and END_STEP contains a set of actions associated with the step. It may contain no actions for a null step. The initial step is identified using the keyword INITIAL_STEP.

The action association is described with the following arguments:

```
ActionName(Qualifier, Timed_Qualifier,
                      indicator_variable );
```

The Time_Qualifier field may be omitted if it is not applicable for the qualifier. The indicator variable is optional.

Examples are:

```
Action_StartUp(L, T#2m, Ready);
Action_Stop(P,Stopped);
```

The transition definition may have one or more steps after the FROM keyword. More than one step indicates that the transition is being used to terminate a number of simultaneous sequences. It may also have more than one step after the TO keyword. This indicates that the transition is being used to initiate a number of simultaneous sequences.

For example, the following definition could be used to describe the transition TrGo shown in Figure 8.10:

```
TRANSITION FROM (Pump1,Pump2,MainGen)
           TO (OnLine, SelectAuto)
   := (GO = 1);
END_TRANSITION
```

The definition of an action is defined by any number of ST or IL statements given between the keywords ACTION and END_ACTION.

8.16 SFC example

We will now consider using SFC to program an industrial lift, the sort of lift that might, for example, be found in a warehousing system.

The lift can be requested to stop at any floor and can be raised or lowered by a motorised winch. When the lift is within a small distance of any floor level, a proximity micro-switch is activated. There are two proximity switches, just above and below the lift stop position. When the lift is approaching a selected floor, the proximity switch signals that the lift should slow down and inch slowly towards the stop position for the selected floor.

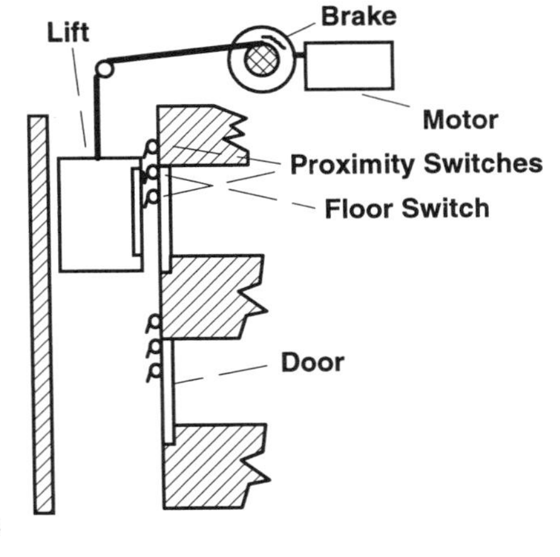

A floor micro-switch is activated when the lift is at the correct level to signal that the lift door can be opened.

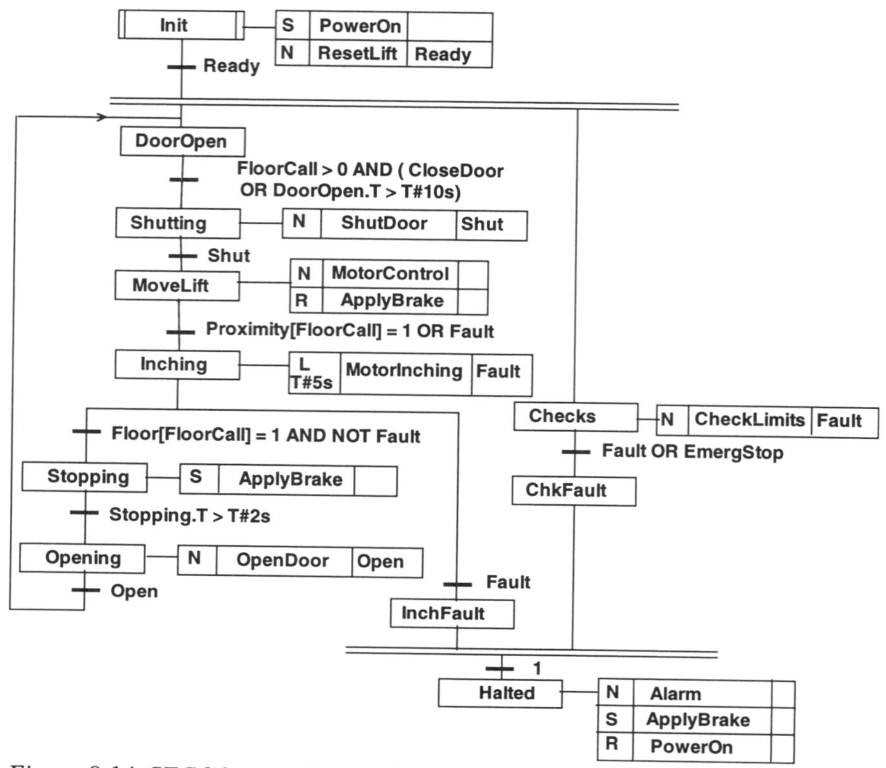

Figure 8.14 SFC lift control example

Part of an SFC to control the lift is shown in Figure 8.14. The initial step 'Init' sets a stored action 'PowerOn' which controls the main power feed to the motor and door mechanism. This action remains active until it is reset as a result of a fault condition in the 'Halted' step.

There are two simultaneous sequences which are initiated by the transition condition 'Ready'; this indicates that the lift has been moved to its starting position and the lift door has been opened.

The main sequence starts with the 'DoorOpen' step. This step remains active until a 'FloorCall' value has been detected and either 10 seconds have elapsed or the 'DoorClose' button has been detected. The 'FloorCall' signal contains the number of the floor where the lift is requested to go next. To keep the example simple, the logic that selects the called floor is not shown but might need to include floor priority, lift request wait times etc.

The 'Shutting' step calls the action 'ShutDoor'; this step remains active until the door microswitch 'Shut' is detected. The 'MoveLift' step resets the stored action 'ApplyBrake' and activates the 'MotorControl' action that is responsible for ramping the winch motor up to speed to move the lift towards the selected floor. When the 'Proximity' switch signal for the called floor is detected, the 'MotorControl' action ceases. Note that the 'Proximity' and 'Floor' switches are held in an array which is indexed by the value of 'FloorCall'.

The 'Inching' step calls an action 'MotorInching' that is time limited, i.e. it can be active for up to 5 seconds. If the 'Floor' microswitch is detected within this time, the 'Stopping' step is activated. This in turn sets the stored action 'ApplyBrake' which controls the winch brake. After 2 seconds, the 'Opening' step calls an action to open the lift door.

Fault detection

The second simultaneous sequence is concerned with monitoring the health of the lift. The action 'CheckLimits' contains logic that monitors such things as, that the lift is moving in the correct direction, is within operating speed limits, and is within lift shaft limits. On detecting an out of limits condition, the action block sets the 'Fault' signal true.

In the main sequence, the 'Fault' signal can cause the 'MoveLift' step to cease immediately, the 'Inching' step to be prematurely deactivated and the 'InchFault' step to be activated.

Notice that if the 'Floor' switch is not detected during the 'MotorInching' action, the 'Fault' signal is set true. This also results in the 'InchFault' step being activated.

If 'InchFault' and 'ChkFault' are both active, the simultaneous sequences both terminate, and the 'Halted' step is activated. This resets the power, activates an alarm and applies the winch brake.

Note: This example is used to demonstrate the SFC language and should not be considered as representative of a control system for an industrial lift where there are issues concerning safety.

Summary

We have seen that Sequential Function Chart provides a flexible and intuitive method for graphically programming sequences of actions in a control system. Some of the main features are:

- SFC is based on established standards and accepted industrial practice.
- SFC allows alternative sequences to be selected using divergent paths.

- There is support for running sequences in parallel using simultaneous sequences.
- Actions and transitions can be programmed using any of the IEC languages.
- A wide range of action qualifiers is provided to give full control over when actions start and stop.
- There is support for hierarchical top down design.
- SFC can be used at any design level, at the top level within a program or at a low level within a function block.
- A textual form is provided to enable SFC designs to be ported to different PLC systems.

Chapter 9

Function blocks

The use of function blocks is one of the most important concepts formalised by the IEC 1131-3 standard. Using function blocks in the design of control software has many benefits including improving software quality and productivity, as reviewed in the introduction - see Chapter 1, Section 9.

In this chapter we will consider:

- The definition of function block types;
- Re-use of function blocks using function block instances;
- Definition and use of IEC standard function blocks;
- A review of some of the example function blocks.

9.1 Function blocks as software building blocks

The IEC 1131-3 standard encourages the development of well-structured software that can be designed using either a top-down or bottom-up approach. An important concept that underpins this objective is the use of function blocks.

A function block is a well packaged element of software that can be re-used in different parts of an application or even in different projects. A function block can provide a software solution to a small problem, or to the control of a major unit of plant.

A function block describes the behaviour of data, the data structure and an external interface defined as a set of input and output parameters. An important characteristic is that the input and output parameters can only use data types as defined within the standard.

Function blocks are the basic building blocks of a control system and can have algorithms written in any of the IEC languages as described in earlier chapters. Well designed function blocks should have use in a wide range of applications and projects.

IEC 1131-3 defines a small repertoire of fairly rudimentary function blocks. There has been no intention to define a comprehensive range of industrial function blocks. This may be because obtaining agreement on a range of general purpose blocks would have probably held up the publication of the standard.

A function block is a type of program organisation unit (POU) which provides an encapsulated algorithm that transforms data within the function block structure. When a function block executes, it evaluates all its variables, which include input, internal and output variables. During its execution, the algorithm creates new values for both the internal and output variables.

A significant feature of a function block is data persistency. Each block can hold data values between each execution, i.e. it has retained state. In some cases, if variables are defined using the RETAIN attribute as discussed in Chapter 3, Section 11, values may be retained between PLC power cycles.

Function block type definitions

A definition for a function block type has two parts:

(1) a specification of the data structure consisting of input parameters, internal variables and output parameters using textual declarations as given in Chapter 3 'Common elements';

(2) an algorithm expressed using either Structured Text, Function Block Diagram, Ladder Diagram, Instruction List or Sequential Function Chart.

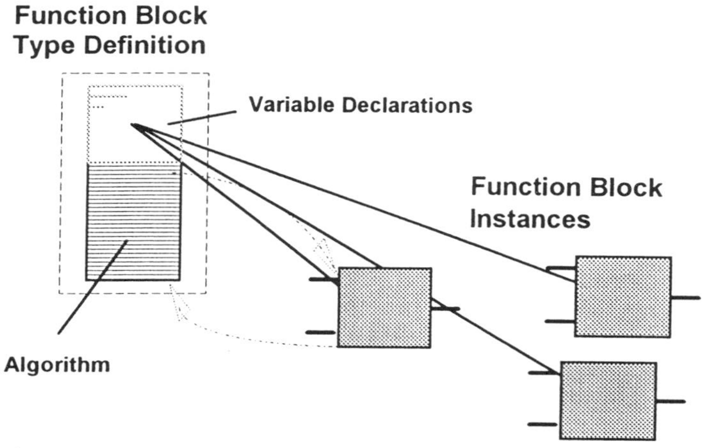

Figure 9.1 Function block types and instances

Function block instances

A function block instance is a particular set of data values held in structures as defined by the function block type. The data in a function block instance can be transformed by the algorithm that is specified in the function block type definition.

The instance only contains data that are unique to a particular application of the function block. This is an important characteristic. Although a function block may have a long and complex algorithm, the memory requirements to hold the variables for each instance of the block may be fairly modest.

Instances of well designed function blocks will generally require less memory than replicas of the equivalent software. The relationship between function block types and instances is shown in Figure 9.1.

Function block example

Consider a DELAY function block, that provides an 'N Sample' delay from the input XIN to output XOUT; this is one of the example blocks defined in the IEC 1131-3 standard. The format for describing the external view of a function block is as shown. Note that inputs are on the left and outputs on the right.

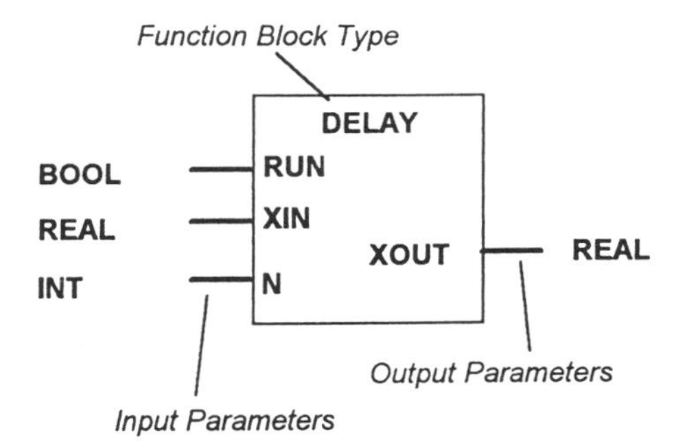

Each parameter is depicted with its name and data type. The function block type is always given just inside the top edge of the block.

The DELAY block is able to delay an input signal by a given number of samples. This is achieved by using an internal array as a circular buffer. Each time the function block executes, a sample is stored at a point in the circular buffer. The output value is obtained by taking a sample from the buffer that is N positions behind the input point.

The signal delay produced by this block is proportional to the function block execution scan rate. For example, a scan rate of 100 ms, and a value of N of 100, will give a 10 second delay.

The definition of the DELAY function block is as follows:

```
FUNCTION_BLOCK DELAY
 (* Variable declarations *)
 VAR_INPUT                (* Input parameters *)
   RUN   : BOOL;          (* Set to 1 to RUN the block,
                             i.e. activate the delay *)
   XIN   : REAL;          (* Input to be delayed *)
   N     : INT;           (* Delay sample width *)
 END_VAR
 VAR_OUTPUT               (* Output parameters *)
   XOUT : REAL;           (* Delayed output *)
 END_VAR
 VAR                      (* Internal Variables *)
   X : ARRAY[0..127] OF REAL; (*Circular buffer *)
   I, IXIN, IXOUT : INT := 0;
 END_VAR

 (* Algorithm *)
 IF RUN THEN
   (* Increment the index for the input *)
   (* save input value in circular buffer *)
   IXIN := MOD(IXIN + 1, 128); X(IXIN) := XIN;
   (* Increment the index for the output *)
   (* save output value in circular buffer *)
   IXOUT := MOD(IXOUT +1,128); XOUT := X(IXOUT);
 ELSE
   (* Reset index values, clear the buffer *)
   XOUT := XIN; IXIN := N; IXOUT :=0;
   FOR I := 0 TO N DO
     X[I] := XIN;
   END_FOR;
 END_IF;
END_FUNCTION_BLOCK
```

The DELAY type definition consists of two distinct parts: (1) variable declarations and (2) an algorithm. The same algorithm is called when any instance of the DELAY block is executed but with a different set of instance variables.

9.2 Function block usage

The following points, which are discussed more fully in Section 3.29, summarise the main points concerning the use of IEC function blocks.

- Externally, it is only possible to access the input and output parameters of a function block instance. Internal variables are not accessible by other program elements declared outside the function block.

  Note: In some IEC 1131-3 implementations, diagnostic facilities may allow internal variables to be monitored and changed.

- A function block instance is only executed if explicitly requested because either (*a*) the function block instance is part of a graphical network of connected blocks which form a program organisation unit or (*b*) it is called in one of the textual languages, i.e. Structured Text or Instruction List.
- Instances of a particular function block type can be used in other function block or program type definitions. An instance of a function block cannot be used within its own type definition - recursion is not allowed.
- Function block instances that are declared as globals using the VAR_GLOBAL construct, are accessible anywhere within a whole resource or configuration.
- It is possible to pass a function block instance as an input to another program, function block or function. This leads to some interesting programming possibilities but can give rise to some unexpected side-effects.
- The current values of function block output parameters can always be accessed in the same way as accessing data for a structure, e.g. Off_timer1.Q.

9.3 Standard function blocks

The IEC 1131-3 standard defines a small range of fairly basic function blocks. However, it is quite surprising how frequently the standard blocks are used when constructing more complex blocks. The symbols for the standard blocks are based closely on the IEC standard 617-12 'Graphical symbols for diagrams'.

Bistables

SR bistable Algorithm as Function Block Diagram

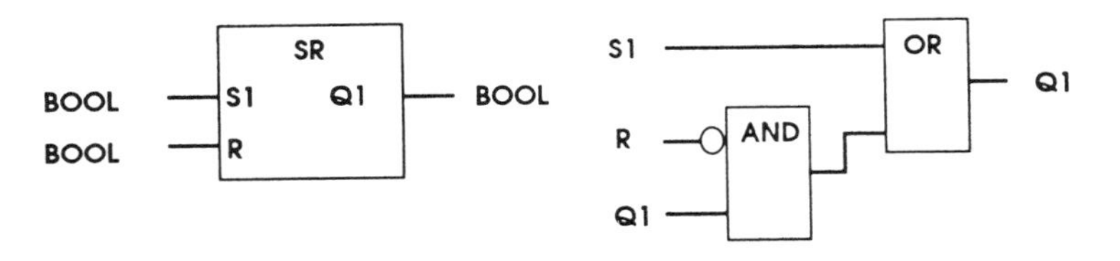

The SR bistable is a latch where the Set dominates. If the Set and Reset signals are both true, the output Q1 will be 1 (true).

RS bistable

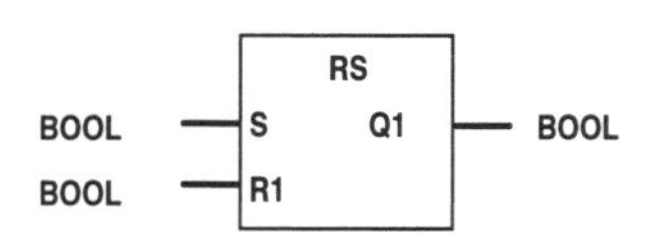

Algorithm as Function Block Diagram

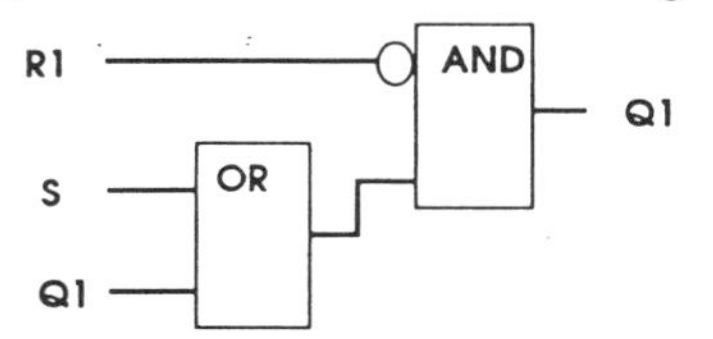

The RS bistable is a latch where the Reset dominates. If the Set and Reset signals are both true, the output Q1 will be 0 (false).

Note: Both SR and RS bistables are defined using Function Block Diagram in which the output value Q1 is used in a feedback path. The use of feedback in graphical languages is discussed in Chapter 5, Section 3.

Semaphore

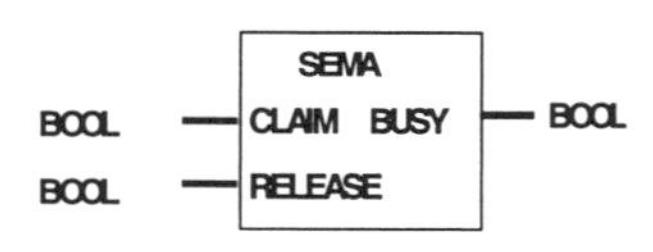

The Semaphore is designed to allow competing tasks to share a particular resource.

A particular Semaphore instance can be claimed by calling the function block with the input CLAIM set to true. If the semaphore is not already in use, the output BUSY is returned as FALSE indicating that the claim was successful.

Algorithm in Structured Text

```
FUNCTION_BLOCK SEMA
   VAR_INPUT
     CLAIM , RELEASE : BOOL;
   END_VAR
   VAR_OUTPUT
      BUSY : BOOL;
   END_VAR
   VAR
       X : BOOL := 0;
   END_VAR
   BUSY := X;
   IF CLAIM THEN
        X:= 1;
   ELSEIF  RELEASE THEN
        BUSY := 0; X := 0;
   END_IF;
END_FUNCTION_BLOCK
```

Subsequent calls of the function block to claim the semaphore will all return with BUSY set to TRUE, indicating that the claim has failed. The semaphore can be released by calling the function block with the RELEASE input true.

The Semaphore function block SEMA provides a mechanism to allow software elements mutually exclusive access to certain resources, i.e. it allows a resource to be shared between two or more software clients.

For example, consider a PRINTER function block that sends messages to a ticket printer. It can only send one message at a time and therefore can only be servicing one task at a time. There may be many other points within different tasks that also require access to the same function block to send messages. The Semaphore can be used to prevent more than one task accessing the PRINTER function block at any one time.

In such situations, it is likely that an instance of the Semaphore function block will need to be declared as global to be accessible from many different function blocks. Whenever any task detects that the Semaphore for the printer is busy, it should wait and attempt to claim the Semaphore some time later.

> Note: It has been said that the IEC 1131-3 definition of the SEMA function block is not an efficient way to control mutually exclusive access to resources. Ideally the Semaphore mechanism should be built into the low level task management of the PLC. An amendment to the 1993 edition of IEC 1131-3 removes the SEMA function block from the standard.

Edge detection

The two edge detecting function blocks 'Rising edge trigger' R_TRIG and 'Falling edge trigger' F_TRIG are provided to detect the changing state of a boolean signal. The outputs of both blocks produce a single pulse when an edge is detected.

Rising edge detector

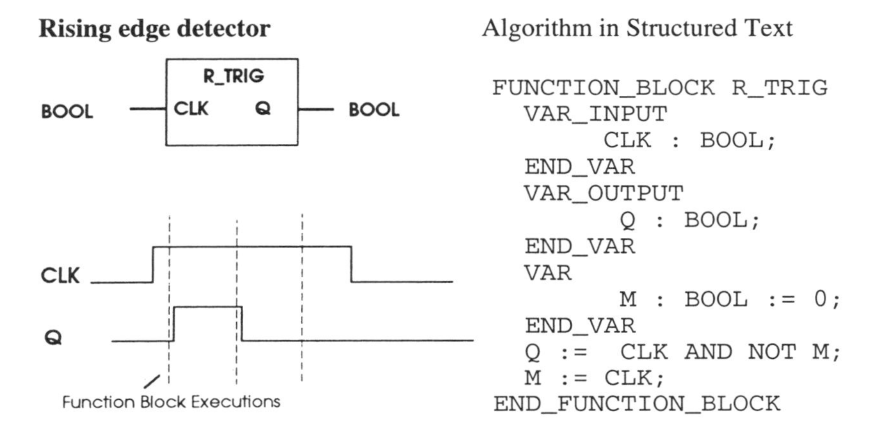

Algorithm in Structured Text

```
FUNCTION_BLOCK R_TRIG
  VAR_INPUT
       CLK : BOOL;
  END_VAR
  VAR_OUTPUT
        Q : BOOL;
  END_VAR
  VAR
        M : BOOL := 0;
  END_VAR
  Q :=  CLK AND NOT M;
  M := CLK;
END_FUNCTION_BLOCK
```

When the CLK input goes true, the output Q is true for one function block execution. The output Q then remains false until a new rising edge is detected.

Falling edge detector

F_TRIG
BOOL — CLK Q — BOOL

CLK
Q
Function Block Executions

Algorithm in Structured Text

```
FUNCTION_BLOCK F_TRIG
   VAR_INPUT
       CLK : BOOL;
   END_VAR
   VAR_OUTPUT
         Q : BOOL;
   END_VAR
   VAR
         M : BOOL := 1;
   END_VAR
   Q :=  NOT CLK
            AND NOT M;
   M := NOT CLK;
END_FUNCTION_BLOCK
```

When the CLK input goes from true to false, the output Q is true for one function block execution. The output Q then remains false until a new falling edge is detected.

Note: These blocks may detect additional edges when there is power-failure or power-recovery. Whether spurious power cycle edges are detected will depend on the implementation and whether the internal state of the edge detection blocks is held in retained memory.

Reserved keywords R_EDGE and F_EDGE can be used as additional attributes of boolean input variables of function blocks and programs. They imply that an instance of the appropriate edge detection function block should be associated with the input.

Edge detecting inputs

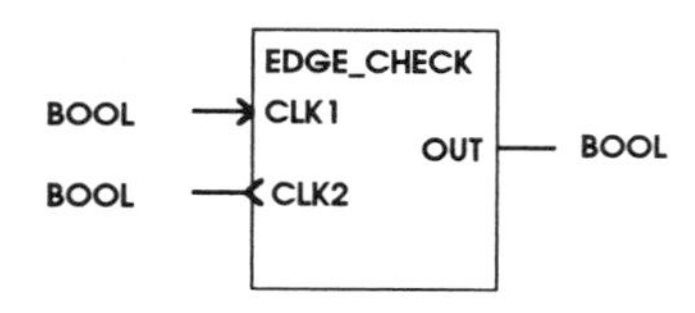

A rising edge input is shown by a left-to-right arrow (see CLK1). A falling edge input is shown by a right-to-left arrow (see CLK2).

Algorithm in Structured Text

```
FUNCTION_BLOCK EDGE_CHECK
    VAR_INPUT
       CLK1 : BOOL R_EDGE;
       CLK2 : BOOL F_EDGE;
    END_VAR
    VAR_OUTPUT
        OUT : BOOL;
    END_VAR
    OUT := CLK1 OR CLK2;
END_FUNCTION_BLOCK
```

In this example, the output OUT is true when either CLK1 detects a rising edge input or CLK2 detects a falling edge.

Counters

A set of general purpose counter function blocks is provided. These are designed to be used in a wide range of applications, for example counting pulses, shaft rotations, completed product batches and so on.

Up-counter

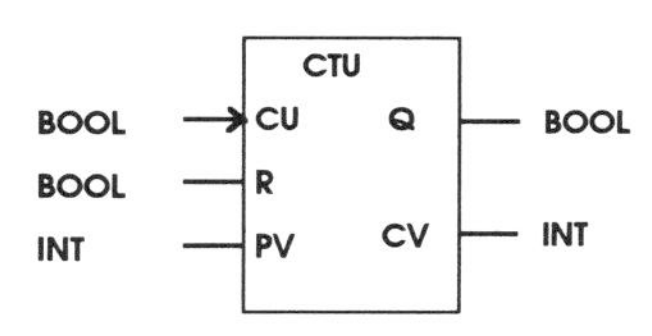

The up-counter block CTU can be used to signal when a count has reached a maximum value.

The CTU block counts the number of 'rising edges' detected at the input CU. PV defines the maximum value for the counter. Each time the function block is called with a new rising edge on CU, the counter output CV is incremented by one.

Algorithm in Structured Text

```
FUNCTION_BLOCK CTU
   VAR_INPUT
     CU : BOOL R_TRIG;
     R : BOOL;
     PV : INT;
   END_VAR
   VAR_OUTPUT
     Q :  BOOL;
     CV:  INT;
   END_VAR
   IF R THEN
     CV := 0;
   ELSIF CU
       AND (CV < PV) THEN
     CV := CV + 1;
   END_IF;
   Q := (CV >= PV );
END_FUNCTION_BLOCK
```

When the counter reaches the PV value, the Q output is set true and the counting stops. The reset input R can be used to set the output Q to false, and clear the count CV to zero.

Down-counter

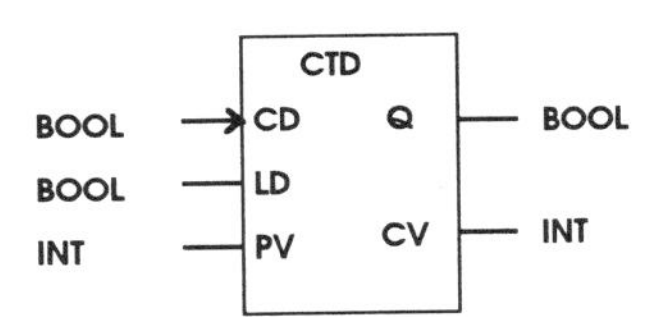

The down-counter block CTD can be used to signal when a count has reached zero, on counting down from a preset value.

Algorithm in Structured Text

```
FUNCTION_BLOCK CTD
    VAR_INPUT
       CD : BOOL R_TRIG;
       LD : BOOL;
       PV : INT;
    END_VAR
    VAR_OUTPUT
        Q :  BOOL;
        CV:  INT;
    END_VAR
```

The CTD block counts down the number of 'rising edges' detected at the input CD. PV defines the starting value for the counter. Each time the function block is called with a new rising edge on CD, the counter output CV is decremented by one.

```
    IF LD THEN
          CV := PV;
    ELSIF CD AND (CV > PV) THEN
          CV := CV - 1;
    END_IF;
    Q := (CV <= 0 );
END_FUNCTION_BLOCK
```

When the counter reaches zero, the Q output is set true and the counting stops. The load signal LD can be used to clear the counter output Q to false, and load the count CV with the preset value PV.

Up-down counter

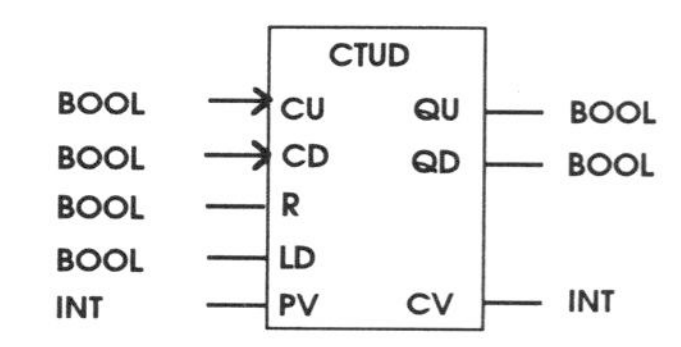

The up-down counter block CTUD has two inputs CU and CD. It can be used to both count up on one input and down on the other.

The CTUD block counts up the number of 'rising edges' detected at the input CU and counts down the 'rising edges' detected on CD. PV defines the maximum value for the counter.

If the counter output CV reaches zero, the QD output is set true and the counting down stops. If CV reaches the maximum value PV, the QU output is set true and the counting up stops.

Algorithm in Structured Text

```
FUNCTION_BLOCK CTUD
   VAR_INPUT
     CU, CD : BOOL R_TRIG;
     R, LD : BOOL;
     PV : INT;
   END_VAR
   VAR_OUTPUT
     QU, QD :  BOOL;
     CV:  INT;
   END_VAR
   IF R THEN
     CV := 0;
   ELSIF LD THEN
     CV := PV;
   ELSIF CU
        AND (CV < PV) THEN
     CV := CV + 1;
   ELSIF CD
        AND (CV > 0) THEN
     CV := CV - 1;
   END_IF;
   QU := (CV >= PV);
   QD := (CV <= 0 );
END_FUNCTION_BLOCK
```

The load signal LD can be used to preset the counter output CV with the value PV. Similarly, the reset signal R can be used to clear the counter output to zero.

The up-down counter could be used, for example, to count items placed on and taken off a conveyor belt. The count value would indicate how many items are on the belt at any time.

> Note: If any of the counter function blocks are executed with count inputs, such as CU, continually held true, they will not detect any rising edges and therefore will not count!

Timers

The standard defines a small set of standard timer function blocks. These can be used in a wide range of applications, for example measuring the time to add reactants to a reactor vessel, eliminating switch bounce, timing mechanical interlocks etc.

The timer blocks are generally used in the graphical languages where there are simple timing requirements. For complex timing operations involving sequencing, it is recommended that Sequential Function Charts and actions with timing qualifiers are used: see Chapter 8, Section 6.

The functionality of all the timer function blocks is implemented typically as part of PLC system firmware, i.e. as system routines - because the timers are closely connected with the operating system timing.

Pulse timer

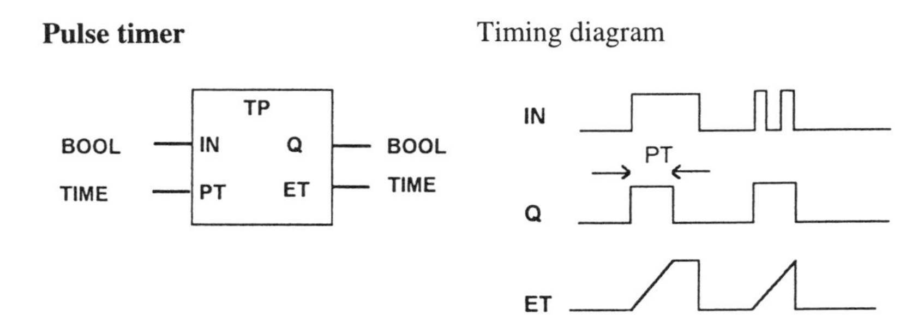

The pulse timer block can be used to generate output pulses of a given time duration.

As input IN goes true, the output Q follows and remains true for the pulse duration as specified by time input PT. While the pulse output is true, the elapsed time ET is increased. On the termination of the pulse, the elapsed time is held until the beginning of the next pulse, at which point it is reset.

The output Q will remain true until the pulse time has elapsed, irrespective of the state of the input IN.

On-delay timer Timing diagram

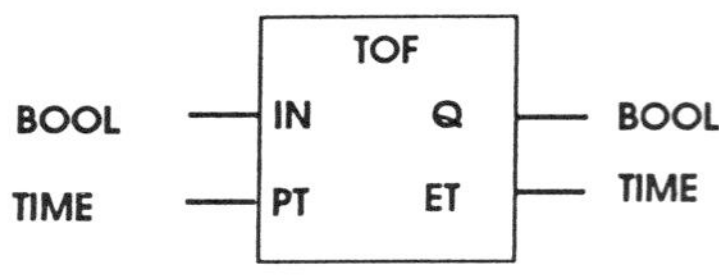

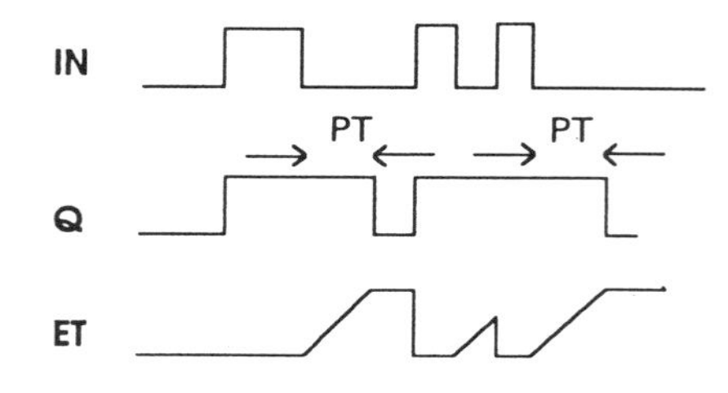

Note: The on-delay timer can be represented alternatively using 'T---0' in place of the text 'TON' in the FBD and LD graphical languages.

The on-delay timer can be used to delay setting an output true, for a fixed period after an input becomes true.

As input IN goes true, the elapsed time at output ET starts to increase. When the elapsed time reaches the time specified by the input PT, the output goes true, and the elapsed time is held.

The output Q remains true until the input IN goes false. If the input is not true longer than the delay time specified in PT, the output remains false.

Off-delay timer Timing diagram

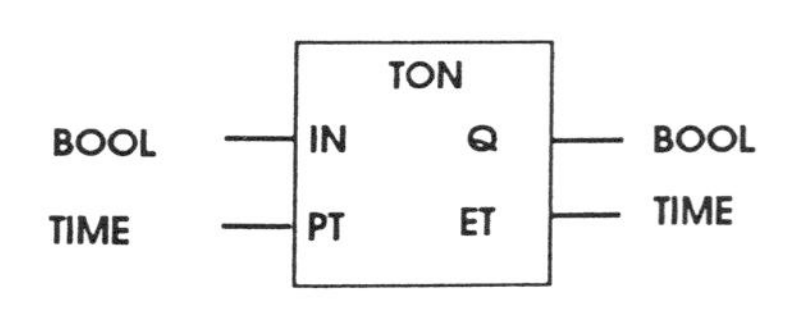

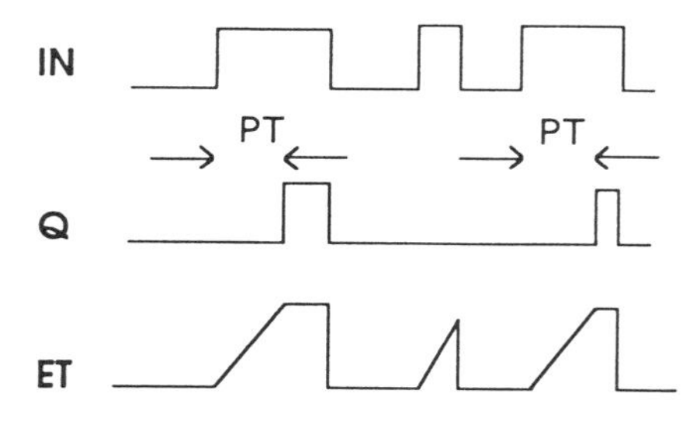

Note: The off-delay timer can be represented alternatively using '0---T'in place of the text 'TOF'in the FBD and LD graphical languages.

The off-delay timer can be used to delay setting an output false, for a fixed period after an input goes false, i.e. the output is held on for a given period longer than the input.

As input IN goes true, the output Q follows and remains true until the input IN is false for the period specified in input PT. As input IN goes false, the elapsed time ET starts to increase. It continues to increase until it reaches the period given by PT, at which point the output is set false and the elapsed time is held.

If the input is only false for a period shorter than the input PT, the output remains true.

A typical use of an off-delay timer is in removing short glitches from noisy input signals, say, from mechanical sensors. An input has to be off for a specified time before its change of state is detected at the timer output.

Real time clock

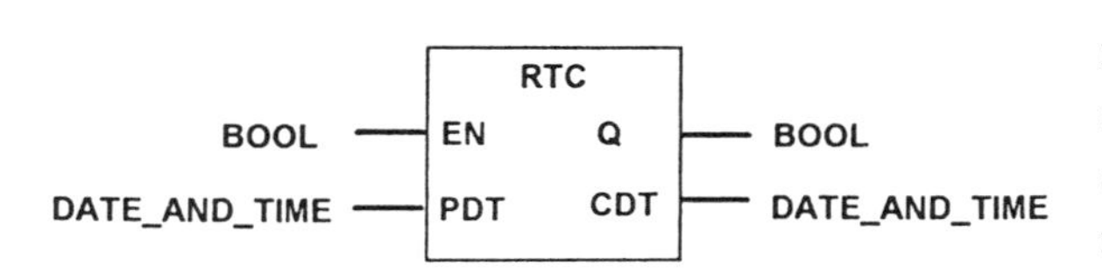

The real time clock has many uses including time stamping, setting dates and times of day in batch reports, in alarm messages and so on.

The behaviour of the real time clock function block is unusual because of the need to set the initial clock value.

The RTC function block can be set to the current time by calling the function block with the input RUN set to true (1), provided that either:

(*a*) the function block has been previously executed with the input RUN set to false (0), or
(*b*) the function block is being executed for the first time.

The input parameter PDT (Preset Date and Time) should be set to the initial date and time. Thereafter, the current date and time can be read by calling the RTC function block with the input RUN set to 1; the value of input PDT is then ignored. The value of CDT does not change until the RTC function block is called again.

The value of the output CDT is not defined if the RTC function block is called with the input RUN set to 0.

Note: The IEC amendment to the 1993 edition of the 1131-3 standard removes the Real Time Clock (RTC) function block definition from the standard because it has proved to be very difficult to implement.

It is likely that many PLCs will only support one instance of the RTC function block. However, by declaring the RTC function block instance as a global variable, the current date and time can be obtained anywhere within programs and function blocks.

The real time clock generally will need to be set by operator interaction at some point when a system is first put on-line. The current time can then be obtained by executing the RTC function block.

RTC example:

```
CONFIGURATION   SYSTEM_ABC
VAR_GLOBAL
  RTC_Clock1 : RTC;
END_VAR
PROGRAM PROG1
VAR_EXTERNAL
  RTC_Clock1 : RTC;
END_VAR
...
IF SetClock = True THEN
  (* Set clock *)
  RTC_Clock1(RUN  := 0);
  RTC_Clock1(RUN  := 1,
       PDT := InputTime );
ELSE
  (* Update current time *)
  RTC_Clock1(RUN :=1);
  CurrentTime:= RTC_Clock1.CDT;
...
```

There may be situations where a PLC can support more than one instance of an RTC function block, in which case each RTC instance will behave as a separate clock. This might be useful when dealing with the times and dates across different time zones. For example, one RTC could be set to the local data and time, while a second RTC instance is preset with the date and time for a different time zone.

9.4 Example blocks

We will now look at a few example function blocks. Some of these are described in the IEC 1131-3 standard. They are provided to demonstrate the capabilities of function blocks. For industrial applications it is likely that these function blocks would require significantly more functionality.

Integral

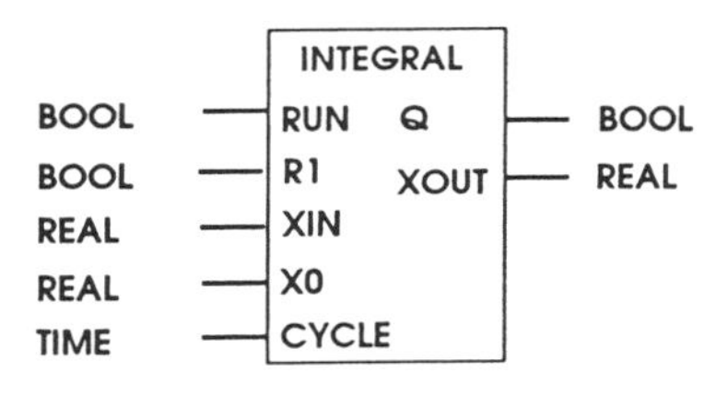

The INTEGRAL function block integrates the value of input XIN over time. The integration can be reset to a preset value X0 by setting the reset input R1 to true.

The CYCLE time defines the time between function block executions. The output XOUT is integrated while the RUN input is true; otherwise the integration value is held.

The Q output is true while the integral is not being reset.

Algorithm in Structured Text

```
FUNCTION_BLOCK INTEGRAL
  VAR_INPUT
   RUN : BOOL; (* Integrate
                  1, Hold 0 *)
   R1 : BOOL; (* Reset *)
   XIN : REAL; (* Input for
                  integral *)
   X0 : REAL; (* Initial
                    value *)
   CYCLE : TIME; (*Cycle
                      time *)
  END_VAR
  VAR_OUTPUT
   Q :  BOOL; (* 1 = Not
                   Reset *)
   XOUT:  REAL; (*Int. Out*)
  END_VAR
  Q := NOT R1;
  IF R1 THEN
     XOUT := X0;
  ELSIF RUN  THEN
     XOUT := XOUT + XIN *
        TIME_TO_REAL(CYCLE);
  END_IF;
END_FUNCTION_BLOCK
```

We will see that the INTEGRAL function block is a useful building block in creating control loop algorithms, such as a PID function block. Other more general uses include, for example, calculating the energy used for an item of plant, such as a pump or fan motor, by integrating power over 'running hours'.

Derivative

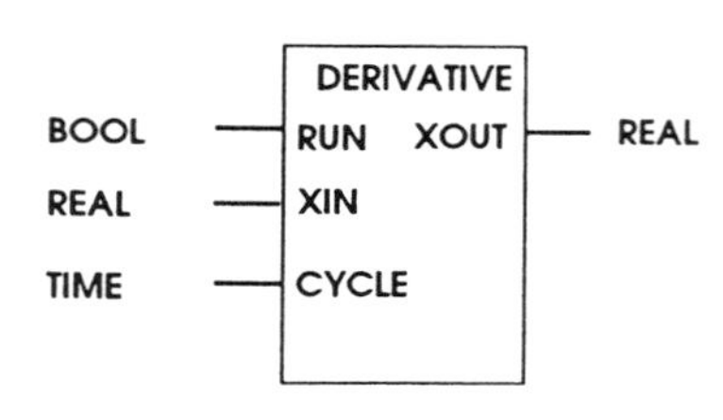

The DERIVATIVE function block produces an output XOUT proportional to the rate of change of the input XIN

The derivative is calculated while the RUN input is true. With RUN set false, the derivative output is reset to zero. The function block CYCLE time is required in calculating the change of input value over time.

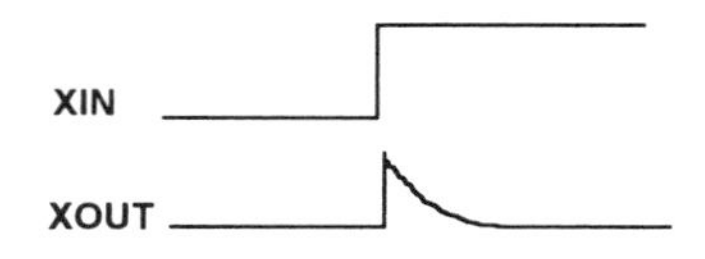

Algorithm in Structured Text

```
FUNCTION_BLOCK DERIVATIVE
  VAR_INPUT
    RUN : BOOL;(*0 =Reset *)
    XIN : REAL;   (* Input
           for derivative *)
    CYCLE : TIME;(*Cycle
           time *)
  END_VAR
  VAR_OUTPUT
    XOUT:REAL;(* Deriv Out*)
  END_VAR
  VAR
     X1, X2, X3 : REAL;
  END_VAR
  IF RUN THEN
     XOUT := ( 3.0 *
      (XIN - X3) +
       X1 - X2 )/(10.0 *
       TIME_TO_REAL(CYCLE));
     X3 := X2; X2 := X1;
     X1 := XIN;
  ELSE
     (* Reset *)
     XOUT := 0.0; X1 := XIN;
     X2 := XIN; X3 := XIN;
  END_IF;
END_FUNCTION_BLOCK
```

The DERIVATIVE function block is mainly used in creating control loop algorithms; it is an important component of the PID function block. It could also have uses in creating alarms or monitoring critical process parameters where the rate of change is important.

PID

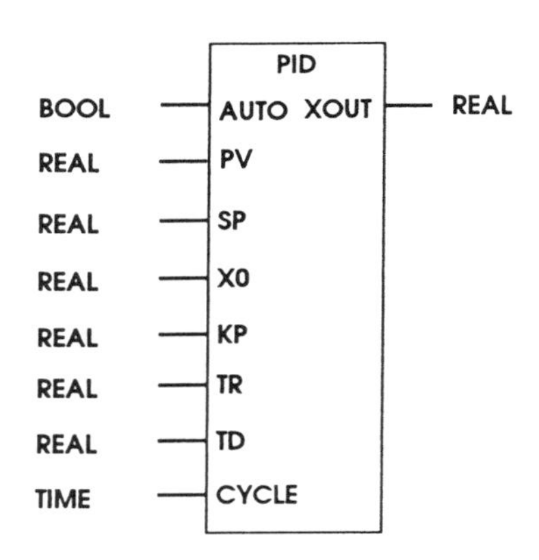

The PID (**p**roportional, **i**ntegral, **d**erivative) function block provides the classical three term controller for closed loop control. It can be used in a wide range of applications where there is a requirement for stable control using feedback of the process value.

The setpoint SP input defines the required value for the process value PV. When the function block is in AUTO it calculates the appropriate value for the output XOUT to move the process closer to the desired setpoint value. The PID algorithm can be tuned to give control with minimum overshoot and steady-state error.

Closed loop control as provided by the PID function block is useful where the process is subject to disturbances caused by external factors or unpredictable disturbances. Examples are when controlling the pressure of a reactor vessel and the pressure changes suddenly due to additional gas being liberated in a chemical reaction, and when controlling a furnace and the internal temperature is affected by changes in the ambient air temperature.

Tuning and the design of PID algorithms are complex topics which are outside the scope of this book. The following figure provides a brief outline of the main features of the algorithm:

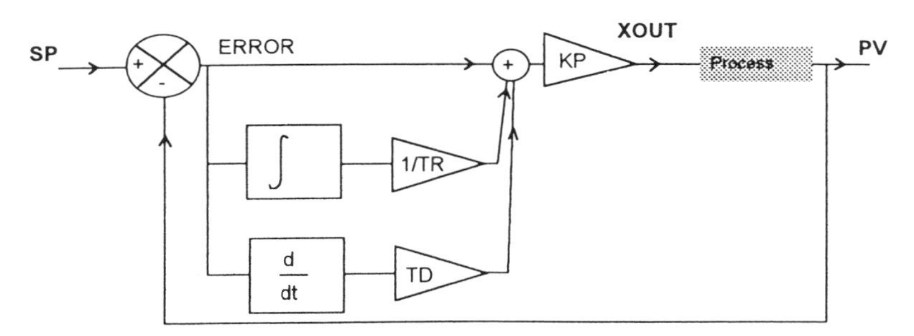

In a typical application, the output XOUT is used to drive an actuator of some form that directly affects the process, e.g. the speed of a pump producing compressed air, the position of a valve delivering gas to a furnace.

The full definition of the example PID block as given in the standard is as follows:

```
FUNCTION_BLOCK PID
  VAR_INPUT
    AUTO : BOOL;           (* 0 - Manual, 1 -Automatic *)
    PV : REAL;             (* Process variable *)
    SP : REAL;             (* Set point *)
    X0 : REAL;             (* Manual output adjustment *)
    KP : REAL;             (* Proportional constant *)
    TR : REAL;             (* Reset time *)
    TD : REAL;             (* Derivative Time constant *)
    CYCLE : TIME;          (* Fun. Blk cycle time *)
  END_VAR
  VAR_OUTPUT
    XOUT : REAL;
  END_VAR
  VAR
    ERROR : REAL;          (* Process error PV-SP *)
    ITERM : INTEGRAL;      (* Integral component *)
    DTERM : DERIVATIVE;      (* Derivative component *)
  END_VAR
  (* PID Algorithm expressed using FBD *)
```

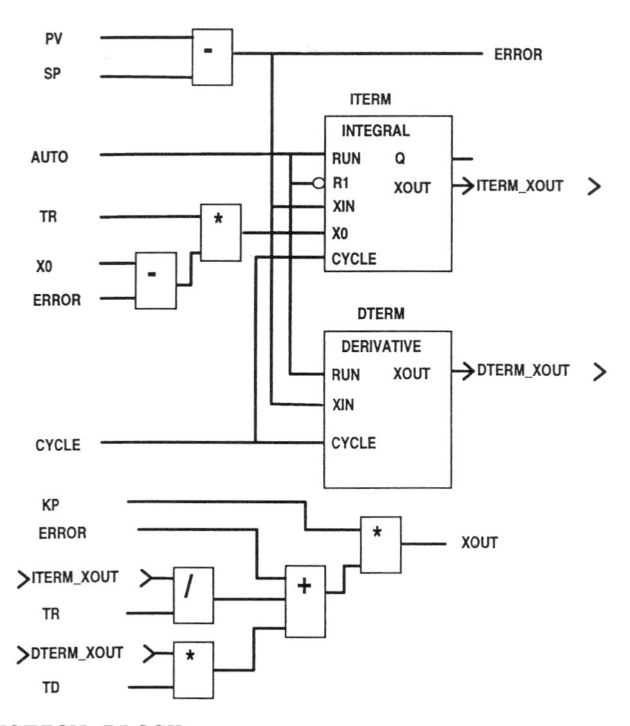


```
END_FUNCTION_BLOCK
```

PID is an example of a composite function block that is built up from using more primitive blocks - in this case using instances of the INTEGRAL and DERIVATIVE function blocks.

Ramp

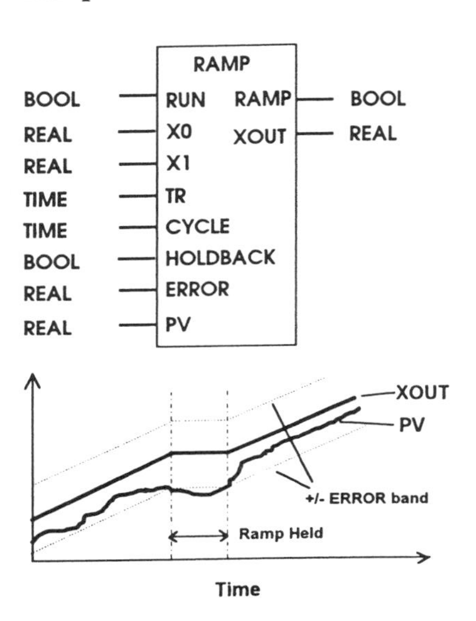

This RAMP function block is modelled on the example given in the standard but with the addition of a 'Holdback' feature. The RAMP block produces an output XOUT that ramps from an initial value X0 to a target value of X1 in a time period given by the input TR.

The ramping function is activated while the RUN input is true; otherwise it is reset.

If the HOLDBACK input is true, the ramping function is held while the absolute error between the output XOUT and the process value input PV exceeds the ERROR value.

A holdback feature is a normal requirement for a process ramp. It is needed to ensure that the process has time to 'catch up' with the ramp value.

For example, consider slowly ramping up the speed of a conveyor belt. For some reason, owing to temporary extra loading on the belt, the speed falls significantly behind the speed demanded by the ramp.

Without holdback, the demanded speed may greatly exceed the actual speed. If the extra loading is removed, there is a danger that the belt will suddenly accelerate in speed to catch up with the ramp speed. Holdback can ensure that the speed never increases dramatically. The ramp is held to ensure that the demanded speed and actual speed never differ by a value greater than specified by the ERROR input.

Note: The ramp can be used both for increasing (ramping-up) and decreasing (ramping-down) values.

The full function block definition is as follows:

```
FUNCTION_BLOCK RAMP
  VAR_INPUT
    RUN : BOOL;          (* Ramp active 1, Reset 0 *)
    X0 : REAL;           (* Initial value *)
    X1 : REAL;           (* Target value *)
    TR : REAL;           (* Ramp duration *)
    CYCLE : TIME;        (* Cycle time *)
    HOLDBACK : BOOL;     (* Holdback when true *)
    ERROR : REAL;        (* Error for holdback *)
    PV : REAL;           (* Process value *)
  END_VAR
  VAR_OUTPUT
     RAMP :  BOOL;       (* 1 = When Ramping *)
     XOUT:  REAL;        (*  Ramping output*)
  END_VAR
  VAR
     XI : REAL;          (* Intermediate ramp value *)
     T : TIME := T#0s; (* Time into ramp *)
  END_VAR

  (* Algorithm *)
  IF RUN THEN

    (* Holdback if PV differs from XOUT by ERROR *)
    IF (HOLDBACK AND ABS(PV - XOUT) < ERROR)
                              OR NOT HOLDBACK THEN
      XOUT := X0 + (X1 - X0) *
           TIME_TO_REAL(T) / TIME_TO_REAL(TR);
      T := T + CYCLE;
      RAMP := TRUE;

    ELSE
      (* Not ramping because of Holdback *)
      RAMP := FALSE;
      XOUT := X0; XI := X0; T:= T#0s;

    END_IF;

  END_IF;

END_FUNCTION_BLOCK
```

Hysteresis

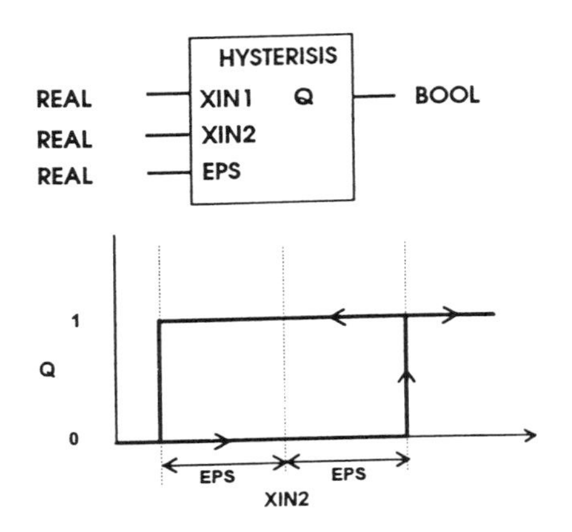

The HYSTERESIS function block provides a hysteresis boolean output driven by the difference of two floating point (REAL) inputs XIN1 and XIN2.

If XIN1 exceeds XIN2 by the value specified by input EPS, the output Q is true. The output Q remains true until the input XIN1 is less than the value XIN2, by the value of EPS. The value of Q then becomes false.

The diagram depicts how the output Q changes as XIN1 increases and decreases.

A hysteresis function block has a wide range of uses. A typical use is in building alarm function blocks where the alarm is triggered by an input exceeding a threshold value. Once the alarm is on, it remains on until the input is less than the threshold value by a specified amount.

The function block type specification is as follows:

```
FUNCTION_BLOCK HYSTERESIS
  VAR_INPUT
     XIN1, XIN2   : REAL;
     EPS : REAL;   (*Hysteresis band +ve and -ve *)
  END_VAR;
  VAR_OUTPUT
    Q : BOOL := 0;(* Initial value is 0 *)
  END_VAR
  IF Q THEN
    IF XIN1 < (XIN2 - EPS) THEN
      Q := 0;   (* XIN1 decreasing *)
    END_IF;
  ELSEIF XIN1 > (XIN2 + EPS) THEN
    Q := 1;     (* XIN1 increasing *)
  END_IF;
END_FUNCTION_BLOCK
```

Ratio monitor

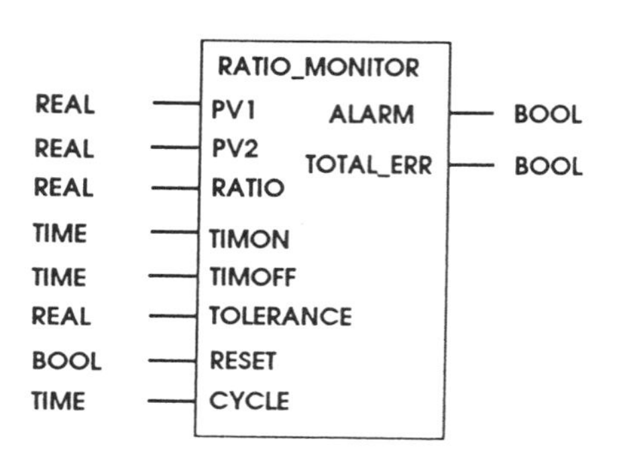

The ratio_monitor function block checks that one process value PV1 is always a given ratio (defined by input RATIO) of a second process value PV2.

If the error between the values of PV1 and PV2, as required to maintain the ratio, goes outside the given TOLERANCE, an alarm timer is started. If the error exists for longer than the time specified by TIMON, the ALARM output is set on.

An off-delay timer is started that ensures that the ALARM stays on for at least a period defined by the TIMOFF input.

The error between the desired ratio value of PV1 and the measured value is accumulated by integrating the ratio error over time. The output TOTAL_ERR defines the integrated error. This value can be cleared using the RESET input.

The ratio_monitor could be used, for example, when feeding two reactants into a process, such as when controlling the gas/air mixture for a gas fired brick kiln. The value of the TOTAL_ERR output gives a quality measure of how well the ratio is being maintained.

The ratio monitor is an example of a composite function block defined using more elementary blocks.

The function block type specification is as follows:

```
FUNCTION_BLOCK RATIO_MONITOR
  VAR_INPUT
     PV1, PV2  : REAL; (*Process values *)
     RATIO : REAL; (*Ratio between PV1 and PV2 *)
     TIMON : TIME; (*Time before Alarm is active*)
     TIMOFF: TIME; (*Time Alarm is held on *)
     TOLERANCE : REAL; (* Acceptable abs. error *)
     RESET : BOOL; (* 1 to clear total error *)
     CYCLE : TIME; (* Fun. block scan time *)
  END_VAR;
  VAR_OUTPUT
    ALARM : BOOL := 0;       (* Ratio error Alarm  *)
    TOTAL_ERR : REAL := 0; (* Integrated error *)
  END_VAR
```

```
    VAR
       ErrorInt : INTEGRAL;(* Error Integrator *)
       TimerOn : TON;      (* Alarm on timer *)
       TimerOff: TOF;      (* Alarm off timer *)
       Error : REAL;
    END_VAR
    (* Check error between process values *)
    Error := ABS( PV1 - RATIO * PV2);

    IF RESET THEN
      (* Reset integrator *)
      ErrorInt( R1 := 1, X0 := 0.0 );
    ELSE
      (*Integrate error *)
      ErrorInt (RUN := 1, R1 := 0,
          XIN := Error, CYCLE := CYCLE );
      TOTAL_ERR := ErrorInt.XOUT;
    END_IF;
    (* Call alarm timers if alarm is active *)
    TimerOn( EN:= Error > TOLERANCE, PT := TIMON);
    TimerOff( EN:= TimerOn.Q, PT := TIMOFF);
    ALARM := TimerOff.Q;
END_FUNCTION_BLOCK
```

9.5 Designing function blocks

When designing a new function block type it is always worth considering all the possible current and future uses of the block. There are many advantages to developing a library of well proven blocks that can be applied to a wide range of problems.

On the other hand, care should be taken not to build so many features into the block that it becomes unwieldy to use.

For more complex blocks consider all the interactions that the block may have with its environment. In particular, consider facilities for:

- Operator control and override
- Commissioning and tuning
- Testing
- Monitoring its internal behaviour

Up to now we have only considered function blocks that are defined using the IEC languages. However, there may be situations where it is necessary to develop a function block algorithm using a language such as 'C' or 'PASCAL'. This normally applies to function blocks that need to interact closely with the PLC operating system.

In some cases, the algorithm may become part of the system software. Provided that the block has input and output parameters that are defined using the IEC data types, it can be used within IEC languages just like any other block.

Summary

The main features of this chapter can be summarised as:

- Function blocks are important software building blocks for constructing control systems using proven solutions.
- Using function blocks speeds software development.
- Function blocks allow PLC memory to be used more efficiently than simply copying identical software.
- IEC 1131-3 defines a small repertoire of elementary, yet very useful, function blocks that includes bistables, timers and counters.
- The standard blocks can be used effectively to construct some very powerful composite function blocks.

In the long term, new types of function block will undoubtedly emerge and become commonplace as the technique of building systems from function blocks gains wider acceptance.

Chapter 10

Programming example

We have now covered all the features of the IEC 1131-3 standard. In this chapter, we will look more closely at the application of IEC 1131-3 to industrial control problems. In particular, the chapter will focus on the following aspects of designing IEC 1131-3 based systems:

- The definition of external interfaces;
- The analysis and breakdown of the control problem into more manageable components, that can be defined as programs and function blocks;
- The partitioning of the control software to run in different PLC resources;
- Selecting the most appropriate programming language;
- Defining the total configuration.

10.1 Steam turbine control example

The example application that we will review in some detail concerns the control of a boiler water feed pump. A water feed pump is used to maintain the water level in the steam drum of a large boiler, as might be found in an oil- or coal-fired power station. The water level in the main boiler must be controlled within safe operating limits.

The level can fluctuate as the steam from the main boiler is used by the power generation turbine or by other services. If the level drops, there is a danger that the boiler riser tubes that absorb heat from the burners will overheat and distort. If the level goes too high, there is a danger that water droplets will pass into the superheated steam supply and damage the main generator turbine blades.

The boiler water feed pump which is driven by a small steam turbine off the main steam supply, pumps de-aerated water (typically re-cycled from condensed steam), under high pressure into the steam drum to maintain the water level.

In this example application, we will consider some of the control aspects of the feed pump steam turbine and the interactions of its control system with the rest of the plant.

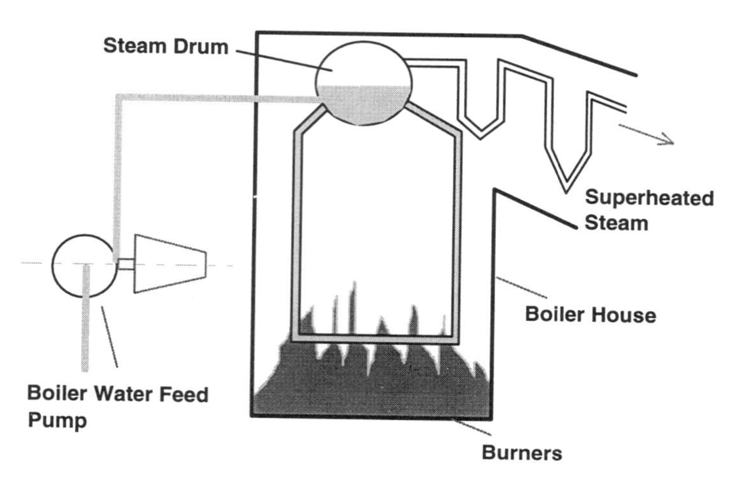

Figure 10.1 Boiler water feed system

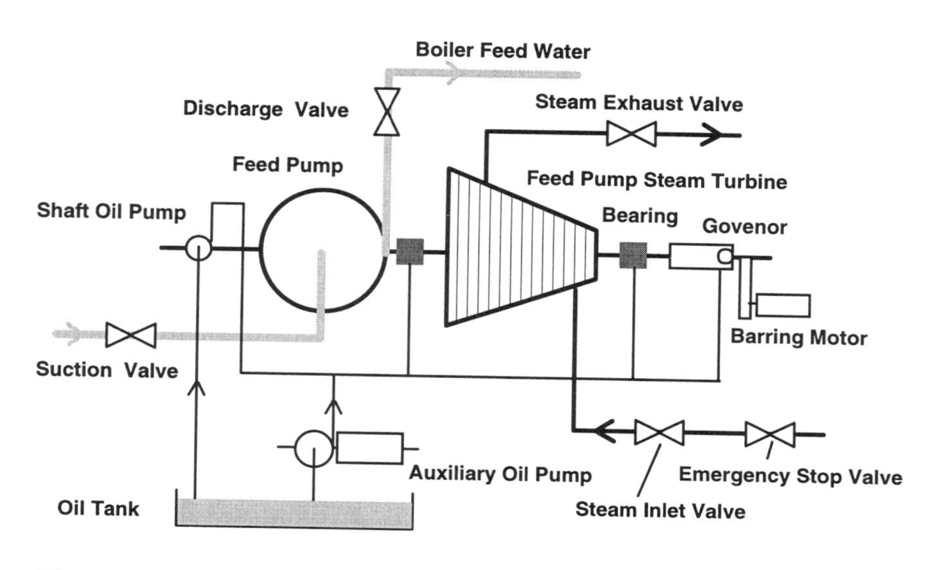

Figure 10.2 Boiler water feed pump

Figure 10.1 depicts the main features of the boiler feed pump. A steam turbine is connected on the same shaft as the feed pump and provides the motive power. It also drives a shaft oil pump that supplies oil for the pump bearings and to power a hydraulic speed governor.

The operator can initiate the pump start-up from the central control room and specify the time for the pump to ramp up to full speed and the initial operating speed. The turbine is ramped up to speed by opening the steam inlet and exhaust valves. To allow the turbine rotor and casing temperatures to stabilise, the turbine is first brought up to an intermediate speed and held at that speed until the casing temperature matches the steam inlet temperature. The steam supply is then increased slowly by opening the steam inlet valve, until the turbine reaches the operating speed.

When the pump is at operating speed, about 3000 rpm, the suction and discharge valves are opened to allow the feed water to be delivered to the steam drum. At this point, the control system switches over to 'modulating control'. A special algorithm is used to regulate the turbine speed according to certain critical process values; these include the drum water level, the feed water supply rate and feed water valve positions. The requested speed signal produced by the modulating control algorithm is used to trim the turbine speed via a hydraulic speed governor.

To cease pumping, which can be triggered by an operator or an emergency shut-down condition, the steam supply is stopped by closing first the steam inlet and then the exhaust valves. As the turbine slows and stops rotating, an electrical barring motor is brought into operation to keep the turbine turning at a slow speed. This ensures that the shaft does not distort as it cools.

An auxiliary electrical oil pump can be brought into operation to maintain oil pressure in the bearings. This can occur whenever the speed of the turbine is too low for the shaft oil pump to be effective, or when the bearing oil pressure is critically low.

10.2 Design approach

The design of the control system for the water feed pump can be considered in the following phases:

1. Identification of the external interfaces to the control system, i.e. definition of all the inputs from the main sensors and outputs to actuators, such as to valves and switchgear.

2. Definition of the main signals exchanged between the control system and the rest of the plant.

3. Definition of all operator interactions, overrides and supervisory data.

4. Analysis of the control problem broken down from the top level into the main control areas; these will then become IEC program organisation units, i.e. IEC programs and function blocks.

5. Definition of any low level function blocks required.

6. Definition of the scan cycle time requirements for the different programs and function blocks.

7. Detailed design of the program and function blocks.

Figure 10.3 shows the main interfaces between the control system and the plant. The primary function is to regulate the speed of the water feed pump to maintain the steam drum water level. However, the complete control system will need to monitor over a hundred input signals that define the state of the plant, including valve position sensors, temperatures, oil pressures etc. There are also numerous outputs for setting steam valves, and controlling auxiliary electrical equipment.

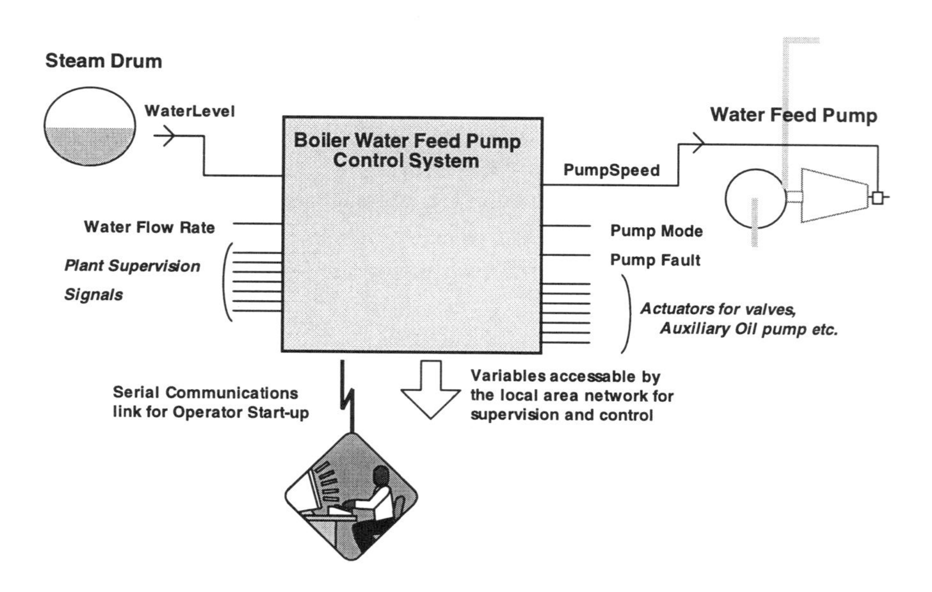

Figure 10.3 Boiler water feed control system

Defining specific data types

In every project it is likely that specific data types are required to represent data in a form that is compatible with the local problem domain. Type definitions should be defined so that they are applicable across the whole project.

Examples:

```
(* Type definition for pump mode *)
TYPE PumpMode : (
  NotAvailable, (* Pump not available
                   for operation *)
  Stopped, (* Pump stopped *)
  Barring, (* Pump rotating at barring speed*)
  Ramping, (* Speed being ramped *)
  Running  (* At operating speed *)
  )
END_TYPE
(* Speed Ramp definition *)
TYPE RampSpec
  Target : REAL;    (* Target speed for ramp *)
  Duration : TIME; (* Ramp duration *)
END_TYPE
```

Definition of plant inputs and outputs

All signals from input sensors concerned with the pump operation, including valve positions, oil pressure sensors, switchgear and temperatures should be defined. Similarly, definitions are required for all the main output signals needed to drive actuators, such as for valve controllers and for operating switchgear, e.g. when powering-up the auxiliary oil pump. Although the main I/O signals will be identified at the start of the design, additional signals may need to be added as the control requirements become clearer.

It is good policy to define a naming convention so that the source and purpose of each signal are clear from the signal name. Each signal should be defined with a data type and should be clearly identified as being a system input or output. In a large project it is likely that a database will be used to hold and access all I/O signal information.

The manner by which signal values are read into or sent out to external devices will vary from PLC to PLC. In this example we have assumed that the PLC hardware is configured so that I/O values are available at pre-defined memory locations.

Example:

```
(* System Inputs *)
P1_Local AT %IX10 : BOOL; (* Local Control *)
P1_PumpSpeed AT %ID50 : REAL;(*Pump revolutions *)
P1_FlowRate AT %ID51 : REAL;(* Water flow rate *)
P1_DrumLevel AT %ID52 : REAL;(* Steam drum level*)
P1_CasingTemp AT %ID53 : REAL;(*Casing Temp *)
...
(* System Outputs *)
P1_SteamIV AT %QD80 : REAL;(*Inlet Valve Pos.*)
P1_SteamEV AT %QX81 : BOOL;(*Exhaust Valve *)
P1_SuctionV AT %QX10 : BOOL;(*Suction Valve *)
P1_DischgV AT %QX11 : BOOL;(*Discharge Valve*)
...
```

The prefix 'P1_' indicates that a signal is associated with pump 1. This is for possible future expansion - a second pump could be controlled using signals prefixed 'P2_'.

Definition of external interfaces

Although this example focuses specifically on the control of the water feed pump, we must not forget that the definition of interfaces with the rest of the plant and other control systems is an important part of the system design. It is likely that the pump control system will have both direct hard-wired signals from other systems and one or more communications interfaces.

The signals and interactions on all of these interfaces need to be defined. Variables that are permitted to be accessed via a plant local area network can be declared using the VAR_ACCESS construct, as discussed in Chapter 2, Section 2.

Example:

```
(* Direct Main Control Inputs *)
P1_StartPermit AT %IX200 :
         BOOL; (* Main Start-up enable *)
P1_PumpTrip AT %IX201 :
         BOOL; (* Emergency Pump Trip *)
...
(* Main Control Outputs *)
P1_PumpMode AT %QW400 :
         PumpMode; (* Pump operation mode *)
P1_PumpFault AT %QX401 :
         BOOL;     (* Pump operating fault*)
...
```

```
(* ACCESS paths for comms network *)

(* Path name to Boiler Feed Pump 1, Speed *)
A_BFP1Speed : P1_PumpSpeed : REAL READ_ONLY;

(* Path name to the Boiler Feed Pump 1, Mode *)
A_BFP1Mode : P1_PumpMode : PumpMode READ_ONLY;
 ...
```

Definition of operator interactions

Interactions with human operators via direct wired switches and indicators can be treated as normal input and output signals.

However, interactions via a communications interface may require the use of special communications function blocks. For example, if a proprietary serial communications interface is being used, it may be necessary to develop a low level function block, say written in 'C'. Handling of serial communications is not covered by IEC 1131-3.

For this example, we have assumed that a low level function block 'Operator_Data_Request' has been developed. This provides an interface with a remote operator panel or display station. It sends a message to the operator and then signals when a response has been returned. The block has two flags, 'Message Ready' and 'Data_Valid', which act as semaphores to signal when the operator message can be updated and when the date entered by the operator is valid.

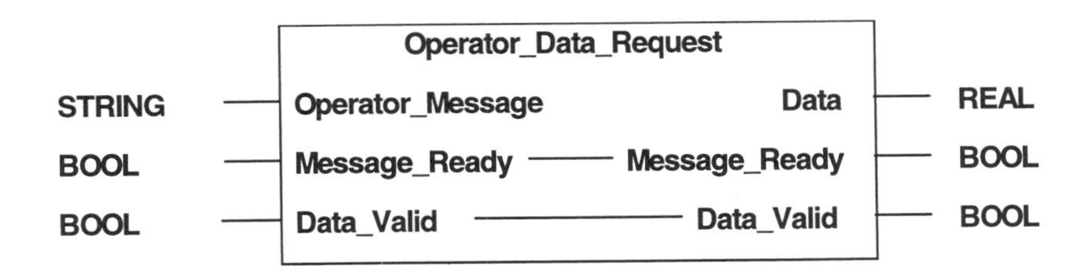

This block is used within the turbine run-up sequence to request the ramp-up speed and time.

Note: Message_Ready and Data_Valid are examples of VAR_IN_OUT variables, i.e. they provide inputs and outputs. The use of VAR_IN_OUT variables is described in Chapter 3, Section 11.

10.3 Control problem decomposition

In this water feed pump example, the principal control functions can be split between (a) the sequencing and auxiliary control, and (b) the modulating control. We have assumed that the PLC provides two processing resources. The Main resource will run the main control sequence for the turbine start-up and control all the auxiliary functions. The second resource, ModulatingControl, is concerned with scanning analogue input signals and running the modulating control algorithm.

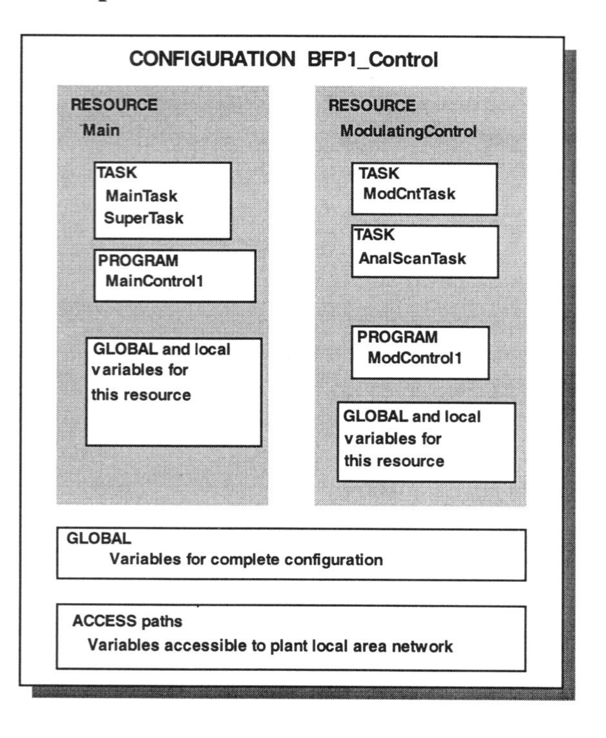

The control software within both resources is encapsulated in program blocks, i.e. MainControl1 and ModControl1. The entire control software is contained in a Configuration BFP1_Control.

Global variables used by software within both resources are defined at the configuration level. These are global variables that either represent inputs and outputs to the system, which are required in both resources, or are globals used to exchange information between the two resources. For example, the ModulatingControl is activated when the pump is revolving at its operating speed. This can be signalled by the MainControl1 program setting a boolean global variable P1_Auto.

Global variables that are only required within a particular resource are declared at the resource level. For example, the constant MaxPumpSpeed is only relevant to the MainControl1 program and therefore is defined as a global variable within the Main resource. Defining the global variables in the resource where they are used can help to simplify the overall design.

Because there is unrestricted access to globals from all lower level software, the number of global variables should be minimised where possible. In this example, all key variables are passed into or out of the programs via formal input and output variables. Although this produces well-structured software, it does result in the main programs having a large number of input/output variables.

Access paths are declared to provide external paths to selected variables. In this case, communications access is restricted to a few principal variables, such as pump speed and operating mode. Additional 'read' access paths to other variables could also be provided for supervisory purposes, such as alarm monitoring. Direct write access to the internal control variables is not provided because of the danger of accidentally disrupting the integrity of the control system.

It is assumed that diagnostic facilities, outside the scope of IEC 1131-3, are available during system commissioning to establish various tuning constants.

Each resource contains one or more task definitions. The task, MainTask, runs the main program block and most of the function blocks within the MainControl resource. A second task, Supertask, is concerned with running a pump supervisory function block. As this block is concerned with monitoring slowly changing inputs such as temperature and vibration levels, it runs at a slower rate.

There is one task in the ModulatingControl resource, ModCntTask, that is responsible for running the modulating control function blocks and function blocks for the signal conditioning of analogue inputs.

Part of the textual definition of this configuration follows. (The characters ... indicate that additional statements have been suppressed for clarity.)

```
CONFIGURATION BFP1_Control
  (* Globals at config. level *)
  VAR_GLOBAL
   (* System Inputs *)
   P1_Local AT %IX10 : BOOL;      (* Local Control *)
   P1_PumpSpeed AT %ID50 : REAL; (* Pump revs/min *)
   P1_FlowRate AT %ID51 : REAL;  (* Water flow rate*)
    ...
   (* Direct Main Control Inputs *)
   P1_StartPermit AT %IX200 :
            BOOL; (* Main Start-up enable *)
   P1_PumpTrip AT %IX201 :
            BOOL; (* Emergency Pump Trip *)
   P1_StartUp AT %IX202 :
            BOOL; (* Start Pump run-up *)
    ...
```

```
    (* System Outputs *)
    P1_SteamIV AT %QD80 : REAL;(*Inlet Valve Pos.*)
    ...
    (* Main Control Outputs *)
    P1_PumpMode AT %QW400 :
              PumpMode; (* Pump operation mode *)
    P1_PumpFault AT %QW401 :
              BOOL;     (* Pump operating fault*)
    ...
    (* Configuration level, shared globals *)
    P1_Auto : BOOL; (* Modulating Control Auto *)
  END_VAR

  RESOURCE MainControl
    VAR_GLOBAL CONSTANT
      (* Define all the global constants *)
      Seq_Scan_Period : TIME := T#100ms;
      MaxPumpSpeed : REAL := 4000;
      ...
    END_VAR
    VAR
      (* Resource level variables *)
      (* System Inputs *)
      P1_AOP_Press AT %ID533 : REAL; (* Aux oil
                                       pump pressure *)
          ...
      (* System Outputs *)
      P1_SteamEV AT %QX81 : BOOL; (*Exhaust Valve *)
      P1_SuctionV AT %QX10 : BOOL;(*Suction Valve *)
      P1_DischgV AT %QX11 : BOOL;(*Discharge Valve*)
      ...
    END_VAR
    (* Task definitions *)
    TASK MainTask (INTERVAL := Seq_Scan_Period,
                         PRIORITY := 5);
    TASK SuperTask (INTERVAL := T#500ms,
                         PRIORITY := 10);
    (* Call the MainControl1 Program *)
    PROGRAM MainControl1 WITH  MainTask :
      MainControl(
        (* List of input variable connections *)
        StartUp := P1_StartUp, (* Pump run-up *)
        Local := P1_Local,  (* Local control mode*)
        PumpSpeed := P1_PumpSpeed,
        FlowRate := P1_FlowRate,
        PumpTrip := P1_PumpTrip,
        ...
        (* List output variable connections *)
        Auto => P1_Auto,
        PumpMode => P1_PumpMode,
        PumpFault => P1_PumpFault,
        SteamIV => P1_SteamIV, (*Inlet valve *)
```

```
            SteamEV => P1_SteamEV, (*Exhaust valve*)
            SuctionV => P1_SuctionV,(*Suction valve*)
            P1_DischgV => P1_DischgV,(*Discharge valve*)
            ...
            (*Assign supervision fun. blk to superTask *)
            PumpSupervision With SuperTask;
          );

     END_RESOURCE
     RESOURCE ModulatingControl
       ...
     END_RESOURCE
     VAR_ACCESS
         (* Variables accessible via comms network *)
       A_BFP1Speed : P1_PumpSpeed : REAL READ_ONLY;
       A_BFP1Mode : P1_PumpMode : PumpMode READ_ONLY;
       A_BFP1Suction :
             MainControl.P1_SuctionV : BOOL READ_ONLY;
       ...
     END_VAR
   END_CONFIGURATION
```

Note: The configuration of a large system may run into many hundreds of lines. However with many IEC implementations, this can be automatically produced by the programming station.

10.4 Program decomposition

We will now look more closely at one of the program blocks for this system. Note that MainControl1 is a program instance of type MainControl. This allows for future system expansion; a second pump could be controlled by simply creating a second program instance, i.e. MainControl2, and linking it to a set of inputs and outputs associated with the second pump.

The program type MainControl consists of a number of function block instances, as depicted in Figure 10.4.

Each function block in Figure 10.4 is concerned with controlling a specific part of the boiler water feed pump. Where possible, all logic concerning the behaviour of a particular part of the plant is contained within a specific function block.

The behaviour of each top level function block within the MainControl program is summarised as follows:

TurbineSequence function block

This block contains the main turbine run-up sequence. In response to the StartUp signal, this block sequences through the various steps to bring the pump up to operating speed. This includes requesting the ramp-up time and target speed from the operator, opening the steam inlet and exhaust valves, holding the ramp at an intermediate speed until the turbine casing temperature has settled, and setting the Auto signal when the pump reaches operating speed. The Auto signal allows the modulating control program to take over the speed control of the turbine.

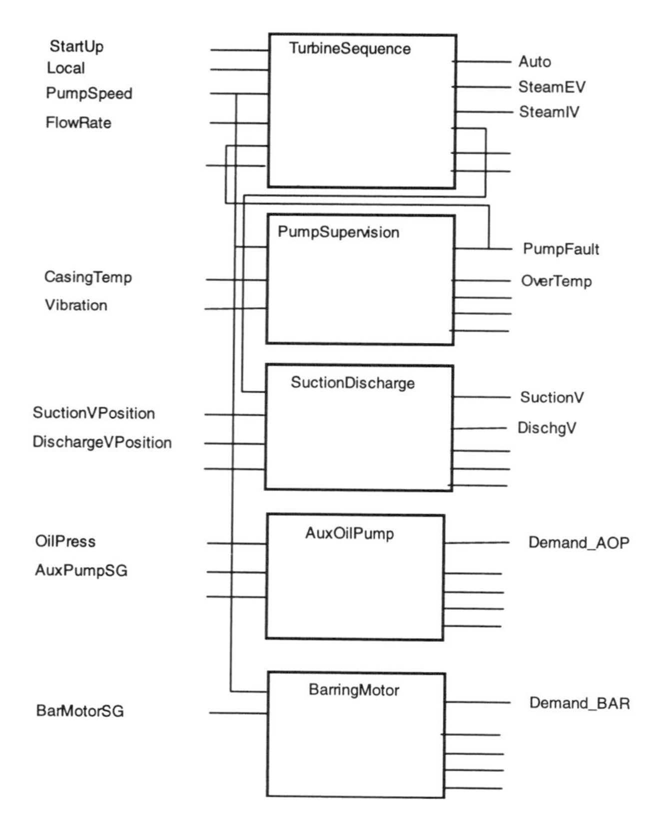

Figure 10.4 Decomposition of the MainControl program

Note: The Function Block Diagram language is ideal for depicting the top level program structure. In this figure, only the primary signals are shown to illustrate the main program structure. Function block input and output names are also not shown to avoid cluttering the diagram. In the full system, each function block may have up to thirty to forty inputs and outputs.

PumpSupervision function block

This block is concerned with checking that the turbine and pump are functioning within normal operating limits. It runs continually and checks that the bearing temperatures, vibration levels and casing temperatures are acceptable. Any discrepancy results in raising the PumpFault output. This is checked by the TurbineSequence, and can halt the turbine speed ramp-up.

SuctionDischarge function block

When the pump speed is near to its operating speed, the water feed suction and discharge valves can be opened. This block ensures that these valves are only opened when the correct conditions exist. It also checks the valve position sensors to ensure that the valves have moved to the requested positions.

AuxOilPump

Maintaining a good oil pressure in the bearings is critical to the health of the pump and turbine. This block continually checks that the oil pressure is always above a critical minimum value. If the pressure falls below a critical level for more than a few seconds, the auxiliary electrical oil pump is immediately powered-up. The block also checks the motor switchgear to ensure that it has operated as requested and that the auxiliary pump, when operating, delivers oil at the expected pressure.

BarringMotor function block

When the pump is taken out of operation, the turbine is allowed to slow down until it stops. At this point a barring motor is engaged to keep the turbine shaft rotating at a slow speed. This function block ensures that the barring motor is engaged and powered-up when requested by a signal from the TurbineSequence function block. The block also ensures that the power to the barring motor is removed when the turbine speed is increased as part of the normal speed ramp-up.

TurbineSequence function block

The TurbineSequence function block is primarily concerned with sequencing through the main steps needed to bring the turbine up to its operating speed. Its behaviour can best be described using a Sequential Function Chart. A part of the turbine start-up sequence is depicted in Figure 10.5.

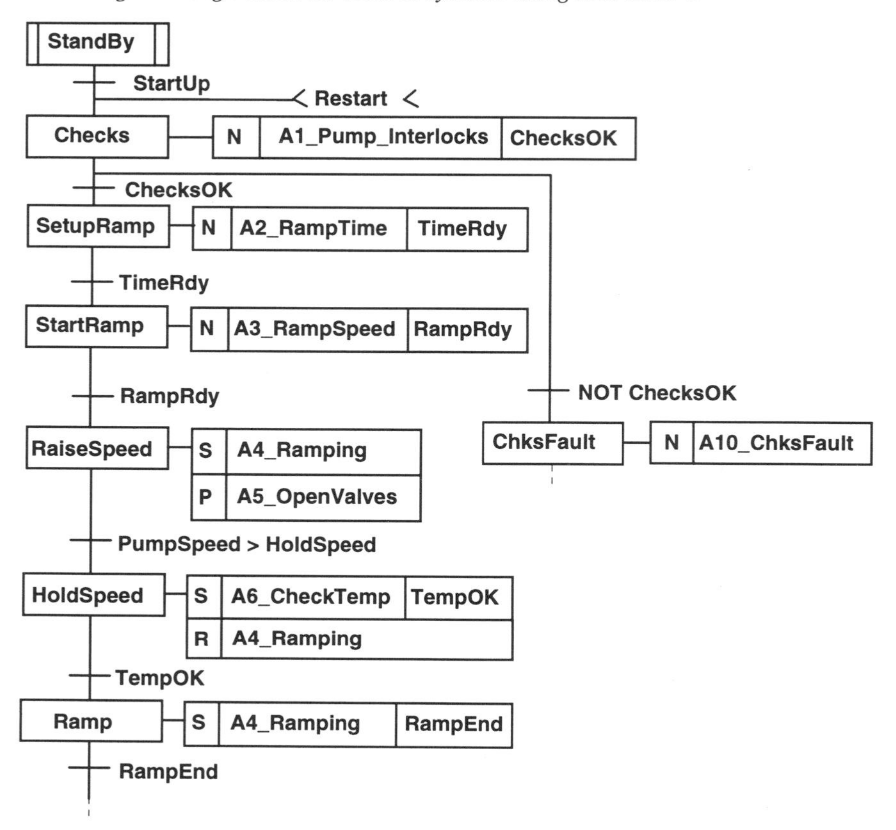

Figure10.5 Turbine sequence SFC

The actions requested by the SFC can be described in any of the IEC languages.

For example, the following Ladder Diagram given in Figure 10.6 describes the logic for the A3_RampSpeed action block. This uses an instance of the Operator_Data_Request function block to prompt the operator to supply the target speed for the turbine run-up. The rising edge of the StartRamp.X signal, that becomes true when the StartRamp step is active, is used to set the message ready signal, Msg_Rdy. This triggers the Operator_Data_Request function block to send the prompt 'Ramp Speed' to the operator.

When the speed value has been received, the function block sets the ramp ready signal, RampRdy, an indicator variable that is tested in the transition that follows the step.

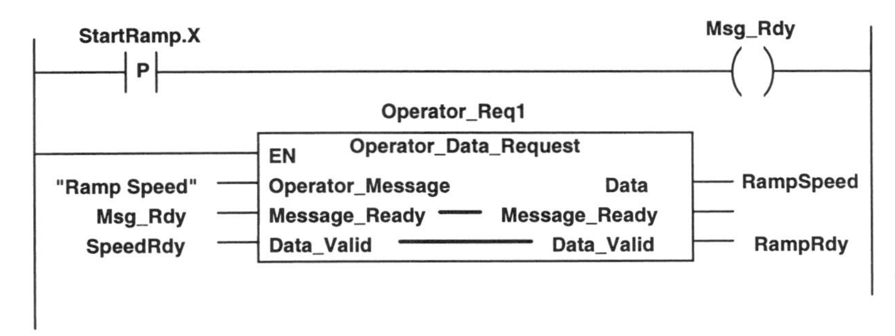

Figure10.6 Action block A3_RampSpeed

As a further example, the following Structured Text statements describe the A5_OpenValves action block.

```
ACTION A5_OpenValves
   StEmergStopCtl(Position := OPEN,Time := T#15s);
     StExhaustCtl(Position := OPEN,Time := T#15s);
END_ACTION
```

This action block calls two function blocks, StEmergStopCtl and StExhaustCtl. These open the steam emergency stop valve and exhaust valves ready for the turbine speed run-up. They are both instances of the valve controller function block, ValveControl, described in a later section.

PumpSupervision function block

This block continually checks that pump and turbine temperatures and vibration measurements are within operating limits. In Figure 10.7 we have shown part of this block described using Function Block Diagram language, but Ladder Diagram or Structured Text would have been equally applicable.

As the block is concerned with monitoring relatively slowly changing inputs, it can be scanned at a lower rate than the other blocks in the MainControl program. In the configuration definition, the PumpSupervision function block is associated with the task called SuperTask, which scans every 500 milliseconds.

Part of the logic for the PumpSupervision function block is depicted in Figure 10.7. The high vibration signals are 'ORed' and passed into an on-delay timer. This ensures that the signals are steady for 5 seconds before the over vibration and pump fault signals are raised.

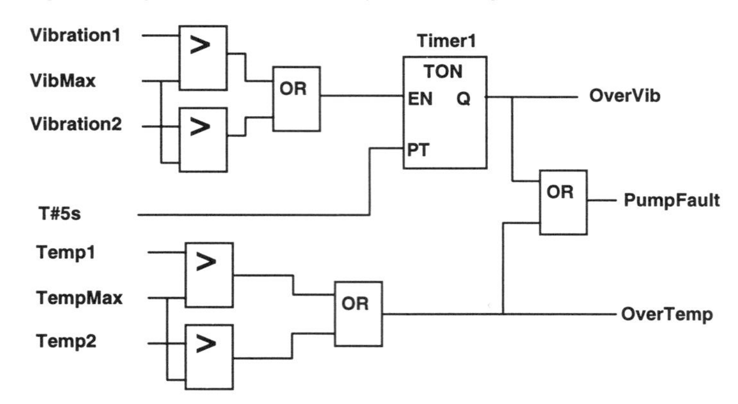

Figure10.7 PumpSupervision function block

10.5 Low level function blocks

So far we have mainly concentrated on the top-down decomposition of the system design into smaller software elements, i.e. into resources, programs and function blocks. However, it is equally important to look for standard software solutions to similar problems that occur in many places throughout the design.

In this example, there are numerous valves to be controlled. Apart from the steam inlet valve which is opened proportionally, the rest of the valves can be controlled using instances of the same valve controller function block.

When the signal from the control system drives a valve actuator to open or close a valve, there is a requirement to check that the valve takes up the required position within a prescribed time. Each valve is fitted with limit switches which indicate when the valve is either fully opened or closed.

The ValveControl function block provides all the logic required to move the valve into the required position, check that the valve is in position and, if not, raise a fault signal indicating that the demanded valve movement has not been detected.

Figure 10.8 shows how the ValveControl function block is linked with the valve actuator. Limit switches on the valve gear feed back the detected valve position.

Note: The logic for this block is very similar to the air damper function block described in Chapter 5, Section 9.

Many other function blocks could be considered for this system, such as a switchgear control block for starting electrical equipment, e.g. auxiliary oil pump and barring motor. With forethought, such blocks can be designed so that they are applicable in other systems.

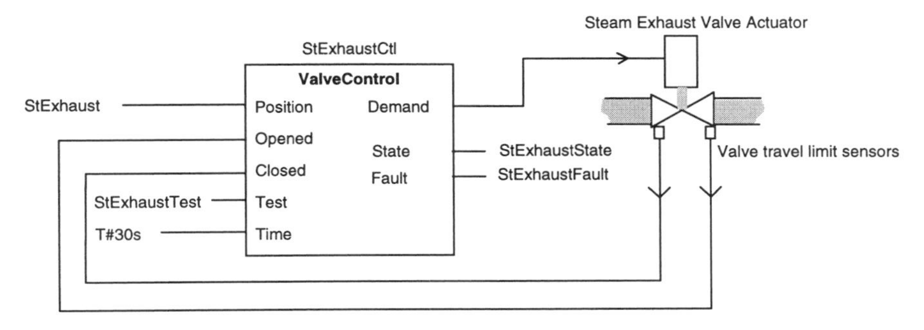

Figure10.8 ValveControl function block

10.6 Signal flow

Because IEC based software can be deeply nested, signal values may need to flow through a number of layers from source to destination. As a result, the value of a particular signal may appear at different levels under different names.

Consider the demand signal for the steam exhaust valve that originates from an instance of the ValveControl function block called StExhaustCtl. This function block exists within the TurbineSequence function block and directly drives the TurbineSequence output SteamEV from its output Demand. In turn, TurbineSequence is an instance within the program MainControl1. The output SteamEV of TurbineSequence directly drives the program output SteamEV.

Finally, the value of the program output SteamEV is written to a resource level variable, P1_SteamEV.

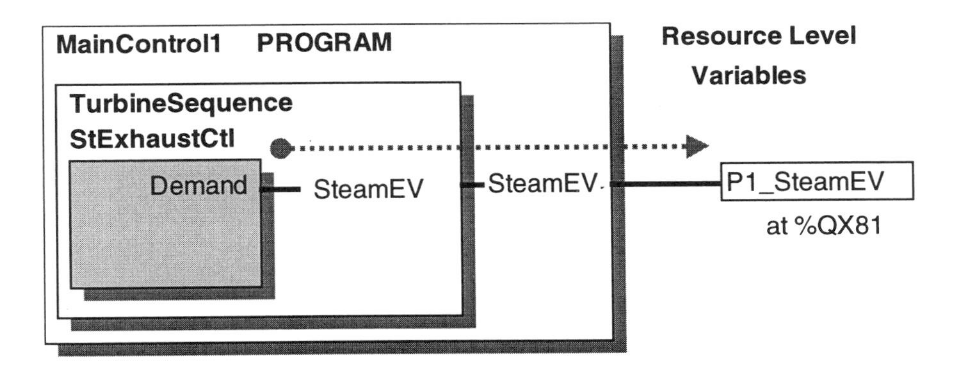

Figure10.9 Signal flow example

The signal flow from function block output StExhaustCtl.Demand through to the resource level variable P1_SteamEV is shown in Figure 10.9.

Similar signal flows exist from input signals through to inputs of internal function blocks.

Although many software layers may exist in a large system, the layers should not normally result in any decrease in system performance. With good compiler design, the transfer of signal values across software block interfaces should not add any significant processing overheads.

> Note: Particular care is required when debugging IEC programs. The context and scope of each variable is always required. For example, the variable SteamEV, as an output of TurbineSequence, should not be confused with the output SteamEV of program MainControl1.

Summary

In this chapter we have reviewed:

- How an industrial control system problem can be broken down into smaller more focused areas.
- How the design can be constructed to run on different PLC processing resources.
- The use of programs and function blocks to construct a hierarchical top-down design.
- The development of re-usable function blocks that can be applied to solve the same problem many times within the same system.
- How different languages can be used to describe function blocks and action blocks.

Chapter 11

Programming station facilities

Producing a programming station that provides all the software management, the graphical and textual editors, the build system and run-time diagnostics required for programming IEC 1131-3 based PLCs is a major software development undertaking.

In this chapter we will review the many aspects of an IEC 1131-3 programming station, particularly:

- Requirements for an intuitive user interface;
- Support for hierarchical design decomposition;
- Library management for re-usable software elements;
- Built-in program consistency checks;
- Provisions for selecting different programming languages;
- On-line diagnostics.

11.1 IEC programming station features overview

Any programming system for control applications based on IEC 1131-3 requires a wide range of features to ensure that programs can be readily developed and maintained by control engineers using a 'user friendly' and robust development environment.

We have seen that IEC 1131-3 style control programs are developed using a top-down design approach where programs generally consist of a hierarchy of graphical diagrams and textual definitions. A programming system should therefore offer facilities that encourage engineers to exploit the benefits of well structured top-down program design.

The main function of a programming system is to provide control engineers and system programmers with facilities to translate their program concepts into a structured and validated set of program diagrams and definitions for which run-time, executable programs can be rapidly created.

Apart from an understanding of the control system languages, engineers using such a programming system should not require any particular or specialist computer skills. Full use should be made of windows style graphical user interface and mouse selection techniques.

The control of all programming facilities should be intuitive through use of well annotated command buttons, menu options or intelligible command icons. Information on using most facilities should be available on-line through help screens or menu descriptions. Where possible a programming system should offer facilities to validate graphical and textual languages.

Automatic checks should be built into the system to validate such things as: that consistent and correct data types are being used, that graphical diagrams are complete and sound, and that textual languages have the correct syntax.

Support for hierarchical design

Where programs are hierarchical, the programming system should provide facilities so that it is possible to 'push-down' on elements in a diagram to view lower level diagrams and program elements.

For example, when viewing a Function Block Diagram, it should be possible to examine the definition behind any element shown on the diagram. This may involve displaying another function block diagram at a lower level in the system hierarchy that depicts the behaviour of a selected function block. If appropriate, the user should be able to close any diagram and return automatically to the parent diagram at a higher level.

Generally, medium to large projects will be developed by a team of engineers. It should therefore be possible to organise and manage all files that constitute a control system as a 'project'.

Multiple users

Facilities are required to allow multiple users to work on a single project simultaneously from different development platforms and then to be able to merge their different program elements into the final system.

Re-use of standard solutions

Through its design, the programming system should promote the re-use of software elements such as IEC programs, function blocks and functions. There should be facilities for creating, storing and retrieving standard elements and for managing software libraries of standard elements.

Over the course of several projects, it is probable that a range of common software solutions will emerge. The ability to turn such solutions rapidly into new standard elements such as functions and function blocks is clearly important in improving software quality and shortening system development times.

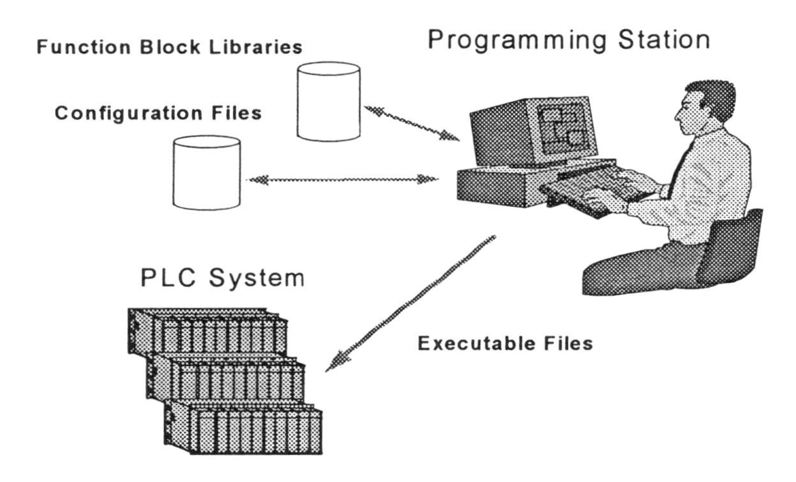

Figure 11.1 IEC 1131-3 programming station

Building the downloadable software executable

When a complete control program has been developed, the programming system should assist in the creation of the run-time executable. Where possible, the system should provide facilities to allow the executable program to be down-loaded into the control system hardware e.g. into non-volatile memory or EEROM.

The status and identity of any down-loaded control program should always be accessible.

Run-time diagnostics

After a control program has been down-loaded, the system should provide on-line diagnostic facilities. It should be possible to display the values of run-time variables, such as the values on function block inputs and outputs. Ideally the values of run-time variables should be displayed in the context of the diagrams used to define the control program.

To support commissioning, full read and write access should be given to run-time variables for system tuning, for making final adjustments to configuration parameters and for calibration of critical sensor measurements.

Access protection facilities should be provided to ensure that all program changes either off or on-line are controlled and authorised using some form of security key, such as a password.

System documentation

Comprehensive documentation of a control software design is important for software validation and maintenance purposes. The programming system should have facilities to print-out all graphical diagrams and text that go together to form a control program.

All print-outs should be organised as a single document with appropriate cross-reference lists to enable the location and use of different software elements to be found rapidly.

11.2 Programming station — feature summary

The following list summarises the main features of a programming system for developing control systems based on IEC 1131-3:

- **User interface** — Ideally a graphical user interface (GUI) should be provided which makes full use of window/icon/mouse techniques with access to all facilities via menus, command buttons or icons.

- **Multiple views** — Programming productivity can be enhanced with facilities that support multiple views of the software (e.g. using windows). It should be possible to view different parts of the software simultaneously so that different parts of a control program can be examined and compared.

  Facilities that allow selected parts of textual and graphical programs to be cut and pasted between views will also help general software productivity. In addition graphical editors should provide zoom and pan facilities so that diagrams may be larger than the physical window size on the computer screen.

- **Intelligent graphical editors** — Graphical editors allow diagrams to be developed in a user-friendly and intuitive manner. For example, graphical lines that define connections between graphical objects should be retained when objects are moved, e.g. when a function block is moved, the lines that connect the block with other blocks should stay attached. This technique is known as rubber-banding.

- **On-line help facilities** — A comprehensive built-in help system should be provided describing all the main facilities including the IEC programming languages, the standard functions and function blocks.

- **Different languages** — It should be possible to select any of the various IEC graphical and textual programming languages for developing different parts of an application for handling control aspects such as: (a) continuous control, (b) interlock logic, (c) alarms, (d) sequencing, (e) interfaces to SCADA and external devices.

- **Data base import/export facilities** — For large projects, a programming station should provide facilities to allow design information to be imported, for example directly from plant diagrams and instrumentation schedules.

  As a minimum requirement, there should at least be an import facility for signal names and I/O descriptions from textual files, databases or other packages such as spreadsheets.

  Similarly, there should be some means to export design information, such as I/O signal names, physical hardware I/O addresses etc. to other databases or to other packages.

- **Hierarchical design** — To support an essential feature of the IEC 1131-3 design philosophy, the programming station should allow the user to break a control system design down hierarchically into smaller more manageable parts, for example into programs and function blocks.

  When using the programming station to examine an existing application, the design decomposition and structure should be easy to understand and explore.

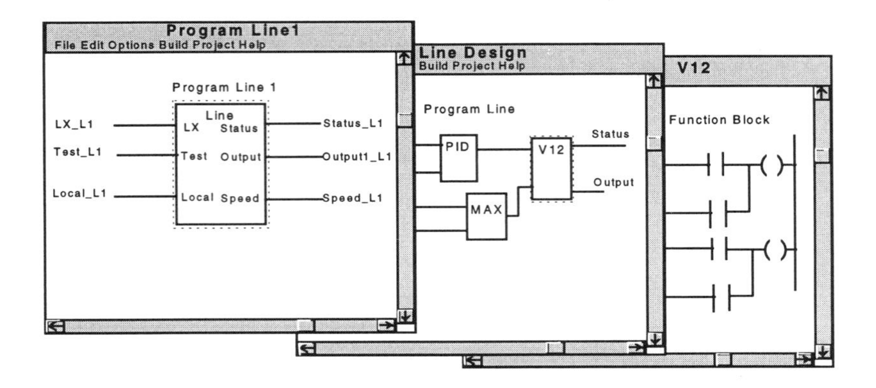

Figure 11.2 Windows showing design decomposition

- **Navigation facilities** — With designs that can be broken down into so many different diagrams and layers, it is important that facilities are provided to enable elements that may be on different diagrams to be found rapidly. This is particularly important when elements, such as function blocks, can be used at different levels in the design hierarchy.

It is also necessary to be able to trace signals from diagram to diagram, for example from a system input, through into the top level program, on into the program level function blocks and so on into the deeper levels of the design.

- **Program validation** — During program development, built-in facilities should be available to validate program elements. Checks should include data type consistency, language syntax and correct diagram composition.

There should also be inter-diagram checks to ensure that inputs and outputs connecting hierarchical diagrams exist and are used correctly, for example that all function block inputs and outputs shown in a top level diagram exist in the associated function block type definition.

- **Library management —** Facilities should be provided for packaging standard software solutions as re-usable elements such as functions and function blocks. Standard elements should be stored in software libraries which are readily accessible. It should be possible to identify and retrieve different revisions of standard software elements as and when necessary.

- **Project management —** With large projects, a programming station should clearly be able to support simultaneous control software development on multiple platforms. This may require support for operation of several programming systems via a network.

  Facilities are also needed to allow control programs to be built by merging source files that have been developed on multiple platforms and to manage sets of program source files as a 'project'. For example, all files associated with a given project should be saved and retrieved as a set of files.

- **Project configuration management** — It is particularly important that facilities are provided to identify the revisions of files as required for a given control program build. Ideally, all the build files should be generated automatically.

- **Documentation —** All information for all sections of a control program, which may be written in different languages, should be available in hard-copy form.

  All hard-copy should be organised to form a single coherent document that describes the complete control program.

  Indexes and cross-references should be available to enable any software element or I/O point to be found easily.

- **Access protection** — Security key protection to prevent program changes by unauthorised personnel should be provided. It is particularly important that on-line changes to a control system are security protected.

- **System build** — The system build process should re-check the consistency of the complete control program. This includes ensuring that all inter-diagram signal connections are consistent, e.g. that all connections have the correct data type and that all function block inputs and outputs that are used in diagrams are actually defined for the associated function block types.

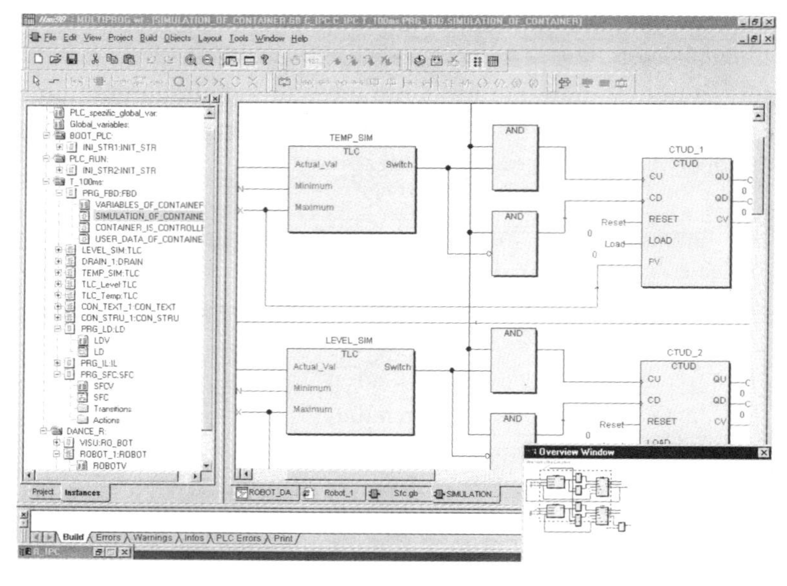

a

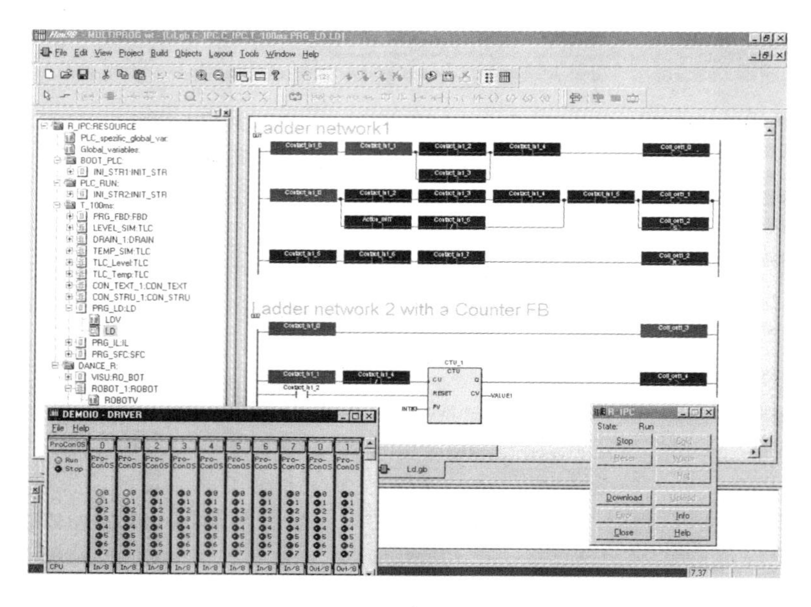

b

Figure 11.3 Example programming system screens

a Function Block Diagram editor

b Ladder Diagram editor

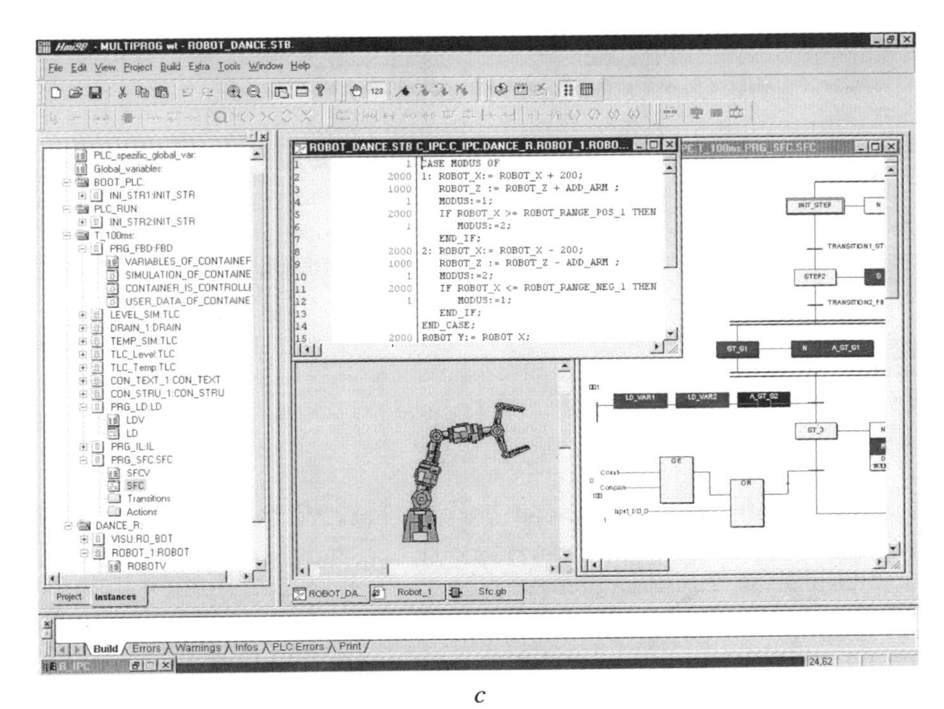

c

Figure 11.3 c Example of using a Sequential Function Chart to position a robot
Screen shots from the MULTIPROG programming system
Courtesy of Kloepper and Wiege Software GmbH

It is also important that PLC address ranges are correct and assigned consistently, e.g. that direct PLC addresses, such as %I101, are not multiply assigned, i.e. assigned to more than one variable.

During the control program build and down-load process, the programming system should ensure that the executable program is configured correctly for the target hardware. For example, the system should check that the built program is consistent with the memory capacity of the target PLC and that the design scan times for different program sections can be achieved.

When coming on-line to the control system, the programming system should be able to identify any executable control programs that are already loaded into the PLC. It should also be able to associate the loaded executable programs with their original source files.

It is important that build information, including the identity and revision of all standard elements such as functions and function blocks, which have been used to create a particular software build, is readily available.

- **On-line diagnostics** — For commissioning industrial control systems it is essential that the programming system provides support for on-line data monitoring. Ideally there should also be options to view graphical programs as they execute showing live signal values, such as at function block inputs and outputs.

With Sequential Function Charts, the programming station should highlight all active steps. If a program is not sequencing through steps as expected, it should be quite clear from the highlighted steps which transitions are holding up the sequence.

- **On-line software modification** — For system tuning, final system configuration and calibration, support for minor in-situ program modification is a useful feature that can save significant time.

However, modifications should be restricted to safe and valid changes. The user should take particular care when making direct changes to a loaded program. It is unlikely that a programming station will be able to check on-line changes with the same rigour as it can check programs built off-line.

All on-line modifications should be recorded for quality audit purposes. If appropriate, the user should have the option to select that changes made to the on-line program also update the associated source files. For example, if control loop tuning parameters are changed via a diagnostic screen, the new values should optionally become part of the control program configuration.

11.3 IEC compliant implementation

A PLC offering an IEC 1131-3 based programming station should come with a comprehensive compliance statement. This should identify all the optional features of the standard that have been implemented.

Within the standard there are a large number of optional language features. Each feature is listed in a table and assigned a feature number.

For example, table 12 in the standard defines data type declaration features; feature 5 is the ability to support structured variables. The PLC may provide or exclude the feature.

A product may either claim to be fully compliant or provide a list of compliance tables in the following format.

Examples:

Table Number	Feature Number	Features description
18	6	Array initialisation
22	1	Functions for integer to real type conversion

The standard states that it is not possible to claim a compliant feature if the feature is implemented using any additional language construct that is not defined in the standard. For example, it is not possible to claim that array initialisation is supported by using another method, such as initialising arrays by directly assigning bits to PLC memory.

The compliance statement should also list all the values of the implementation dependent parameters. These are parameters defined by IEC 1131-3 that may vary between products, generally because of limitations of the PLC hardware or operating system: for example the maximum length of identifiers that can be used, the maximum comment length, the maximum number of array subscripts.

Error detection

The standard defines over twenty different categories of software error condition. These include errors such as:

— numerical result exceeds range for data type,
— invalid character position specified in a string function,
— return from a function without assigning a value.

Ideally, most of the errors in language syntax and usage should be detected by the programming station while the software is being developed. Other errors will only be found when the configuration is being built. Finally there are errors such as 'division by zero in an arithmetic expression', which can only be detected when the software is executing.

The standard requires that the compliance statement includes a detailed description on how each category of error is treated and where it is detected in the software life-cycle.

Summary

In this chapter we have briefly reviewed the features of a programming station suitable for developing software for IEC based PLCs. As the standard has only recently been published, there are few PLC products on the market today that can offer all the features discussed in this chapter.

As it has been estimated that it can take at least twenty-five man-years to develop a full featured IEC based programming station it may be some time before products start to appear in number.

More sophisticated systems will emerge in due course as the cost of computer hardware continues to fall and the use of graphical programming techniques becomes more widespread.

To summarise, the main features that we should come to expect include:

- An interface that is truly 'user friendly';
- Support for formally decomposing complex designs;
- A selection of programming languages;
- Support for re-use of standard solutions;
- Error avoidance and early error detection;
- Powerful, graphical run-time diagnostics.

Chapter 12

Communications

So far we have only considered part 3 of the IEC 1131 PLC standard. As we have seen, IEC 1131-3 addresses all aspects related to programming PLC based control systems. However, it does not address communications between PLCs.

This chapter reviews part 5 of the IEC 1131 standard, which focuses solely on PLC communications. In this chapter we will review:

- The main scope of the IEC 1131-5 standard;
- The communications model;
- The PLC functional model;
- Accessing PLC status information;
- Communications function blocks;
- Mapping of communications function blocks on to MMS.

12.1 Background to IEC 1131 part 5

The only feature defined in the IEC 1131-3 standard that goes any way towards providing inter-PLC communications is the VAR_ACCESS construct. This simply defines access paths to named variables within the PLC application software. IEC 1131-3 does not define how the communications facilities are provided or controlled. The PLC communications model showing the use of the VAR_ACCESS construct is discussed in Chapter 2, Section 2.

Part 5 of the IEC 1131-3 standard is concerned with facilities that can allow PLCs that are connected via a communications network to exchange status and control information and initiate executive commands, such as re-starting resources and downloading configurations.

Note: It is the intention that this chapter provides an overview of the main concepts defined in part 5 of the IEC 1131 standard that deals with PLC communications. For full details see the 1995 edition of IEC 1131-5.

IEC 1131-5 is concerned with all the external communications aspects of a PLC. There are two aspects to communications services; part 5 addresses:

1. **server facilities** — the behaviour of the PLC required to support and respond to external service requests;
2. **client facilities** — the services that a PLC can request from other PLCs.

The standard is intended to be independent of the underlying communications protocols. It therefore assumes that various communications layers, such as the physical layer, data link layer and so on, as defined in the ISO seven layer model, exist or that their equivalent behaviour is provided by the communications subsystems.

IEC 1131-5 effectively exists above the applications layer of the International Standardisation Organisation (ISO) reference model.

The ISO model consists of seven layers that provide the following facilities:

Physical layer: controls the low-level exchange of information bits including transmission rates, and bit encoding over the transmission medium.

Link layer: responsible for the reliable transmission of information units (packets or blocks) and their dispatch to addressed stations on the network.

Network layer: establishes virtual paths between stations in the network, e.g. transmitting a stream of data packets from end to end via switching nodes.

Transport layer: responsible for transporting messages between communications partners. It controls the data flow and ensures that the data are not corrupted and arrive in the correct order.

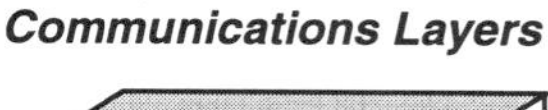

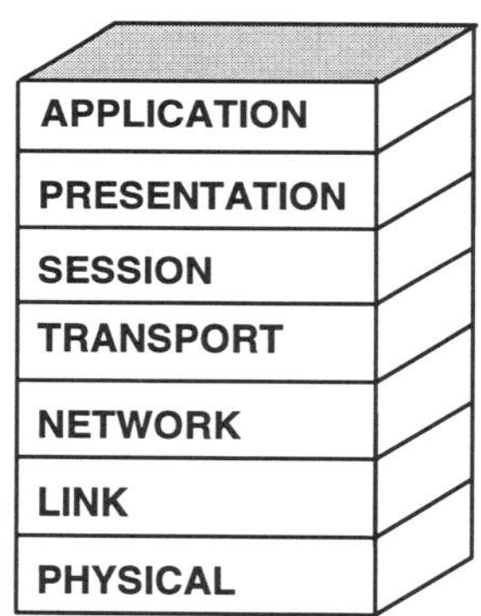

OSI reference model

Session layer: controls the exchange of messages in the transport connection, e.g. establishing a communications session, controlling the re-start after interruption.

Presentation layer: decodes or re-encodes data to ensure that data can be transferred between systems where data encoding is different.

Application layer: a set of protocols, i.e. message transactions to provide specific facilities, e.g. file transfer, database queries etc.

In many industrial networks, many of the ISO layers are either null or collapsed to produce a smaller layered stack. This is generally necessary to improve the communications response time and throughput.

The nature of these underlying communications layers is all hidden by facilities defined within IEC 1131-5. This theoretically allows PLCs to communicate over any type of network. Of course, this will depend on providing the appropriate software to map the facilities required by IEC 1131-5 on to the communications subsystem.

For example, IEC 1131-5 compliant PLCs may be able to communicate via MAP[1] based networks using the Manufacturing Message Specification (MMS) applications layer, via Ethernet and TCP/IP or via a Fieldbus based protocol.

Communications facilities defined in IEC 1131-5 are developed from concepts derived from the languages standard, IEC 1131-3; these are namely access paths and communications function blocks.

Communications function blocks and associated data types are defined using concepts and languages defined in IEC 1131-3.

12.2 Communications model

The main features of the top-level communications model defined by the IEC 1131-5 standard are depicted in Figure 12.1. A PLC may behave as a server, by providing information and responding to service requests, or as a client, requesting information and initiating service requests. Other devices, such as a supervisory system or other non-IEC 1131-3 compliant devices, may also act as clients and request information and services from a PLC server.

Communications protocols, such as Ethernet, will generally allow an indefinite number of server and client PLCs to co-exist on the same network. Note that, in some cases, a PLC may behave as a server to some PLCs and as a client to others.

The IEC 1131-5 standard only defines the communications facilities within PLCs, e.g. PLCs A and B as shown in Figure 12.1. It does not cover the behaviour within external client devices.

---

[1] *MAP, an industrial communications standard initiated by General Motors, is discussed in Chapter 1, Section 3.*

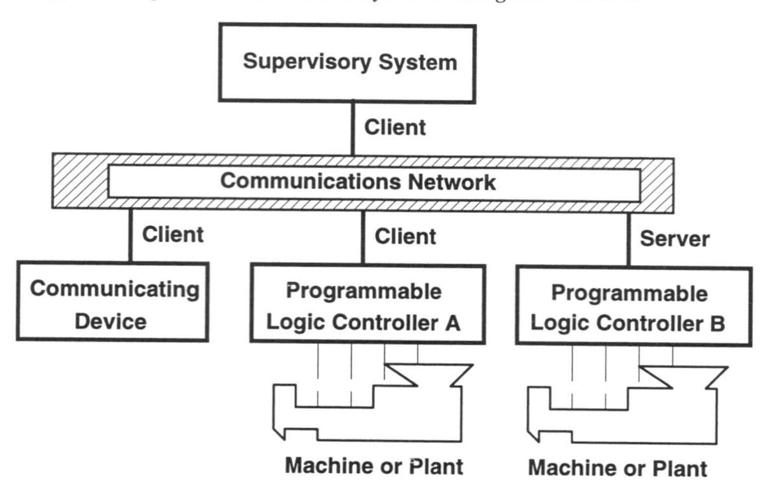

Figure 12.1 IEC 1131-5 communications model

The IEC 1131-5 standard states that each PLC may optionally provide communications facilities to support the following application-specific functions:

- Device verification
- Data acquisition
- Control
- Program execution and I/O control
- Application program transfer
- Synchronisation between user applications
- Alarm reporting
- Connection management

12.3 PLC functional model

Figure 12.2 depicts the main functional aspects of a PLC. The executable program defined using the IEC 1131-3 languages requires the support of a number of subsystems. Remote PLCs, devices and supervisory systems all require communications services to monitor their health and status and, in some cases, control each of these functional subsystems.

For example, a supervisory system may require access to the program diagnostics for analysing a program fault. A remote PLC may need to monitor the

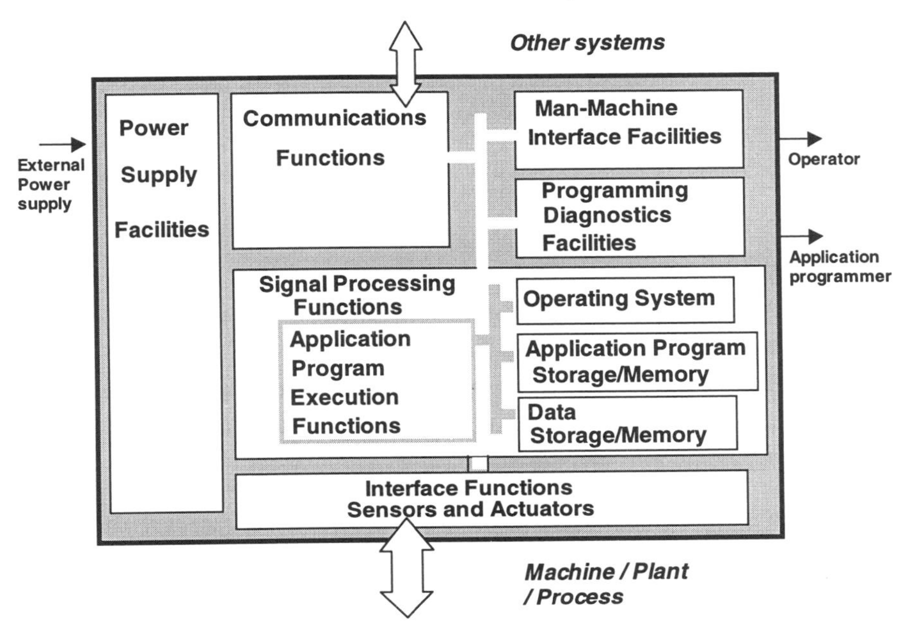

Figure 12.2 PLC functional model

health of the PLC hardware, e.g. to check that the I/O scanning, memory and power supplies are fully operational.

12.4 PLC status information

The IEC 1131-5 standard states that status information is provided for each of the main subsystems as listed in Table 12.1.

Table 12.1 PLC Subsystems

No.	Subsystem
1	PLC, complete system
2	I/O subsystem
3	Processing unit
4	Power supply subsystem
5	Memory subsystem
6	Communications interface
7	Implementation specific subsystems
	...

For each subsystem, the status information will be presented as a data structure having a standard format and defined using IEC 1131-3 data types. For example, the status information for the complete PLC includes the items listed in Table 12.2.

Table 12.2 Main PLC summary status

	Item	Description	
1	**Health**	**Good**	All subsystems are fully operational.
2		**Warning**	At least one subsystem is in the warning state; no subsystem has the bad state.
3		**Bad**	At least on subsystem has the bad state.
4	**Running**	If true, at least one part of the application software is running and controlling some aspects of the PLC's behaviour.	
5	**Local control**	If true, the PLC is under some form of local control, e.g. by the operator. This may limit the network access or disable the PLC's reception of remote executive commands.	
6	**No outputs disabled**	If true, indicates that the PLC has full control over all outputs. If one or more outputs have been forced to take particular values, e.g. while debugging, then this item will be false.	
7	**No inputs disabled**	If true, this indicates that the PLC can access the real values of all inputs. If this is not true, it indicates that one or more input values cannot be accessed, e.g. as may occur in the early stages of commissioning.	
8	**I/O forced**	If true, indicates that either (a) one or more inputs are being forced by the PLC's local programming and diagnostic system to take particular values or (b) the values of one or more outputs have been overridden by the local programming and diagnostic system.	

	Item	Description
9	**I/O subsystem**	If true, it indicates that the 'WARNING' or 'BAD' state is associated with an I/O subsystem.
10	**Processing unit subsystem**	If true, this attribute indicates that 'WARNING' or 'BAD' is caused by a processing unit subsystem.
11	**Power supply subsystem**	If true, this attribute indicates that 'WARNING' or 'BAD' is caused by a power supply subsystem.
12	**Memory subsystem**	If true, this attribute indicates that 'WARNING' or 'BAD' is caused by a memory subsystem.
13	**Communications subsystem**	If true, this attribute indicates that 'WARNING' or 'BAD' is caused by a communications subsystem.
14	**Implementer specified subsystem**	If true, this attribute indicates that 'WARNING' or 'BAD' is caused by an implementer-specified subsystem.

Each subsystem within the PLC has status information containing a similar set of items. The status of each subsystem (and any subsystems contained therein) has an initial item called 'health', which can take one of three states:

GOOD — The subsystem has not detected any faults — it is fully operational.

WARNING — The subsystem has detected one or more faults, which may limit its ability to completely perform its intended function. For example, an I/O subsystem may occasionally be reading erroneous values from certain inputs.

BAD — The subsystem has detected one or more faults which implies that the subsystem is unable to perform its intended function.

The status information for each subsystem may also supply product specific status details, such as additional error diagnostics, operational states, and local operator key states.

Each set of status items will be defined as an IEC data structure. IEC 1131-5 states that such status information shall be accessible via pre-defined access paths or from directly represented variables at defined fixed addresses. The standard defines a complex set of reserved direct addresses for accessing the status of each subsystem within a PLC.

For example, %S0 is the direct address for the main PLC state; %SS3 is the direct address of the state of subsystem 3.

12.5 Communications function blocks

We will now review the set of communications function blocks that are defined by the IEC 1131-5 standard to allow IEC compliant PLCs to exchange information and control signals.

Table 12.3 lists the facilities provided by the standard communications function blocks.

Table 12.3 Communications function blocks

Communications Facility	Communications Function Block
Connection management	CONNECT
Device verification	STATUS, USTATUS
Data acquisition	READ, USEND, URCV
Control	WRITE, SEND, RCV
Alarm report	NOTIFY, ALARM
Variable scope management	REMOTE_VAR

Connection management

Communications function blocks read and write to remote PLCs via communications channels. A channel can be established by calling an instance of the CONNECT function block and supplying it with the full network address of a remote PLC. The CONNECT block returns a local identity of a channel. Subsequently, all communications function blocks which need to communicate with the particular remote PLC can use the open channel.

Device verification

The STATUS and USTATUS blocks provide facilities to read the status of remote PLCs. These would typically be used to ensure that a remote PLC is fully operational at the start and at regular intervals during a joint control operation.

Data acquisition

This concerns reading the values of variables from remote PLCs. The standard supports two methods:

1. **Polled**: The READ function block can be used to read the values of selected variables, either periodically or on a specific trigger condition.

2. **Programmed**: The data messages are supplied on a time or condition determined by the remote PLC. The USEND block in a remote PLC can be used to send unsolicited data which are received locally by the URCV block.

Control

Two methods are provided for the interaction of local and remote control software:

1. **Parametric**: This allows a local PLC to modify the behaviour of a remote PLC by writing values to key variables. The WRITE function block allows values to be written to selected 'access path' variables in a remote PLC.

2. **Interlocked**: This provides a control transaction where a local PLC requests that a remote PLC performs an operation and then signals back that the operation is complete. This can be achieved using the SEND function block in the local PLC (client) and the RCV block in the remote PLC (server).

Alarm report

A PLC can signal selected remote PLCs when certain predetermined alarm conditions arise. Remote PLCs can then send an acknowledgement back to the local PLC to indicate that the alarm has been received. The ALARM and NOTIFY function blocks can be used to generate acknowledged and unacknowledged alarms, respectively.

Function blocks outline description

The following descriptions provide an overview of each block. In the IEC 1131-5 standard the behaviour of each block is formally defined using timing and state transition diagrams.

CONNECT

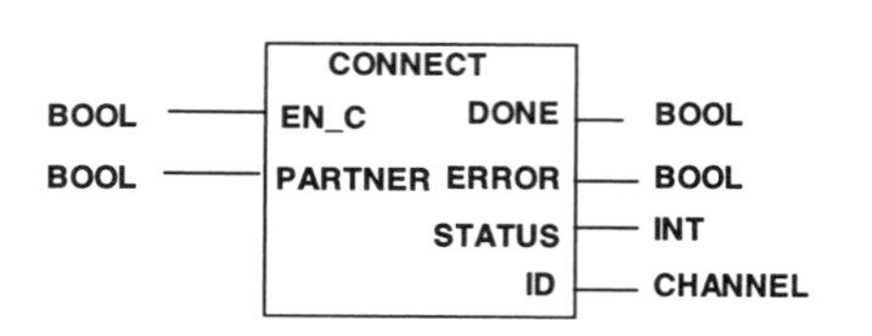

This block is used to establish a communications channel to a remote PLC partner. The remote PLC should have a unique address (e.g. node address) on the network.

The channel ID can be used by other function blocks when communicating with the remote PLC. The EN_C input is an enable that should be true to make the channel connection.

Note: The data type of PARTNER is implementation specific.

STATUS

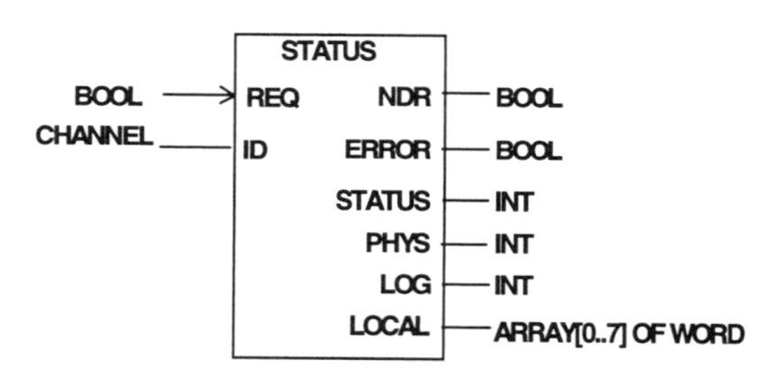

The STATUS block provides remote status information for device verification.

The status information is only requested on the rising edge of the REQ input. PHYS defines the physical status, LOG the logical status, including software status, and LOCAL contains additional implementation specific status.

Physical status refers to the PLC hardware; logical status concerns the status of the application software.

The input ID should specify the channel identity as established by the CONNECT block. NDR is set if a new user data message is received, i.e. if the status information has changed. If the ERROR output is false, the other outputs provide status information from the remote PLC.

USTATUS

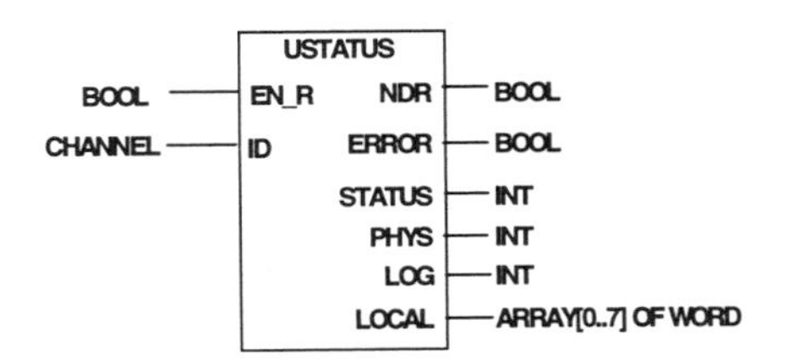

The USTATUS block provides status information from a remote PLC. While USTATUS enable input EN_R is true, the status information is updated whenever the status of the remote device changes. PHYS defines the physical status, LOG the logical status, including software status, and LOCAL contains additional implementation specific status.

READ

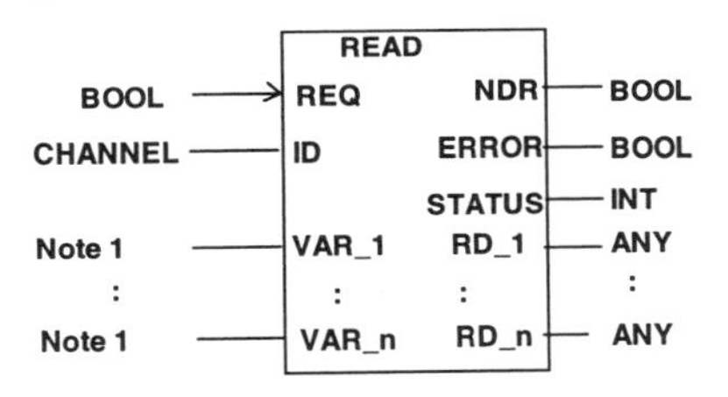

Note 1: Can be VAR_ADDR data type;
see the REMOTE_VAR function block.

The READ block polls a remote PLC for values of one or more variables. A list of variables can be requested by supplying names as inputs. After a short delay due to the network communications, the values of the selected variables are presented on the function block outputs.

The READ block only fetches new values for the selected variables on the rising edge of the REQ input.

WRITE

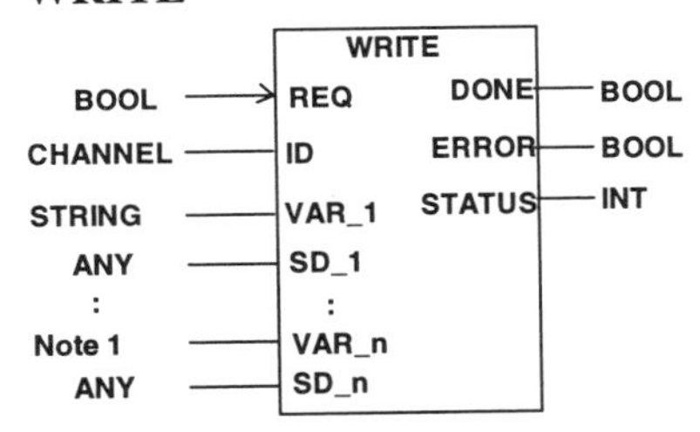

The WRITE block writes one or more values to one or more variables within a remote PLC. A list of variable names specify access variables in the remote PLC. The DONE output becomes true when the remote write operation has completed.

Note 1: Can be VAR_ADDR data type;
see the REMOTE_VAR function block.

Note 2: Inputs and outputs of the READ and WRITE blocks concerned with remote variables use the generic data type ANY, implying that the data type will change to match the type of variable being read or written. The use of data type ANY is discussed in Chapter 3, Section 7.

The READ and WRITE function blocks use extensible inputs and outputs, i.e. the number of inputs and outputs can be changed to suit the set of variables being addressed; see Chapter 3, Section 19.

USEND

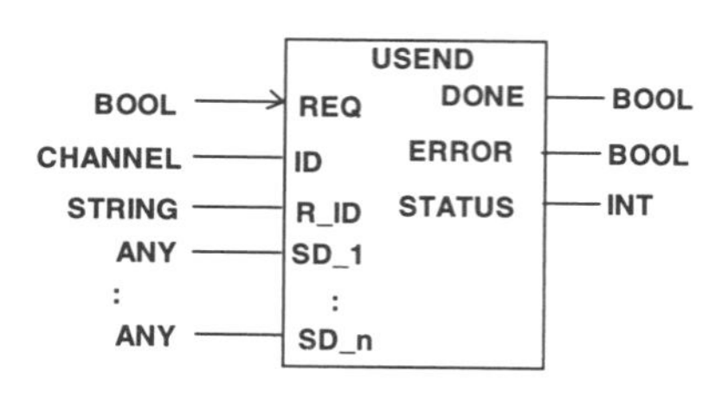

The USEND function is used to transmit data to a particular instance of a URCV function block existing in a remote PLC. The name of the remote URCV instance should be supplied at the R_ID input. The number of SD_n inputs is extensible.

URCV

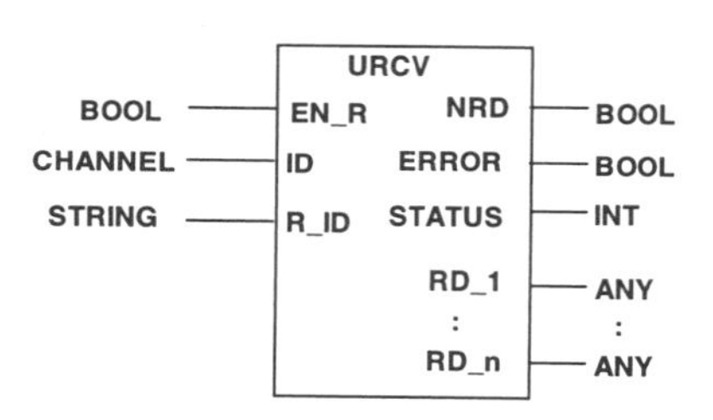

The URCV block receives data for one or more variables from an instance of a USEND block in a remote PLC. The R_ID input is used to specify the instance name of the remote USEND block. The number of RD_n outputs is extensible.

Note that the number of variables and the variable data types of the SD_i inputs of the USEND block should match the RD_i outputs of the URCV block. Figure 12.3 depicts an example of USEND and URCV blocks being used to transmit job details.

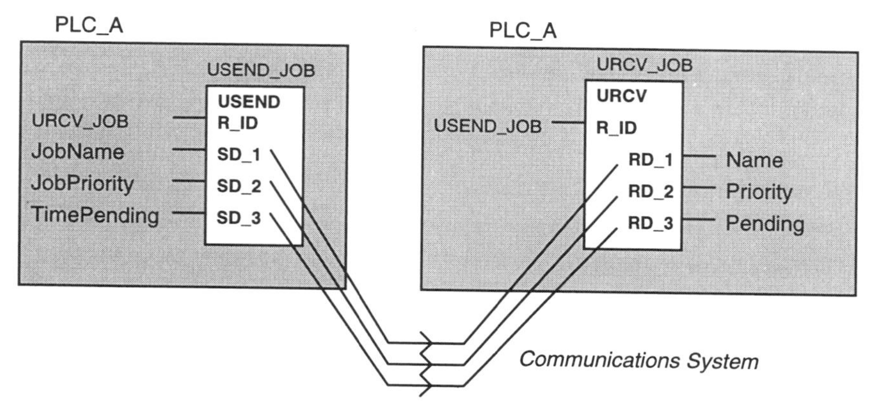

Figure 12.3 Using USEND and URCV function blocks

SEND

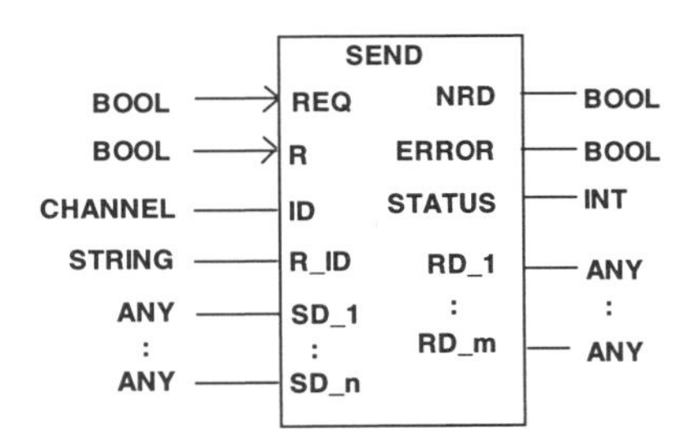

The SEND block provides an interlocked data exchange with an RCV block in a remote PLC. The SEND block transmits the values of one or more variables to a remote RCV block corresponding to the channel ID established by the CONNECT block. The number of SD_n inputs is extensible.

The name of the remote RCV block should be specified at input R_ID. On receiving the data values, the remote PLC then returns a set of values as a response to the SEND block. These then appear at outputs RD_1 to RD_m.

RCV

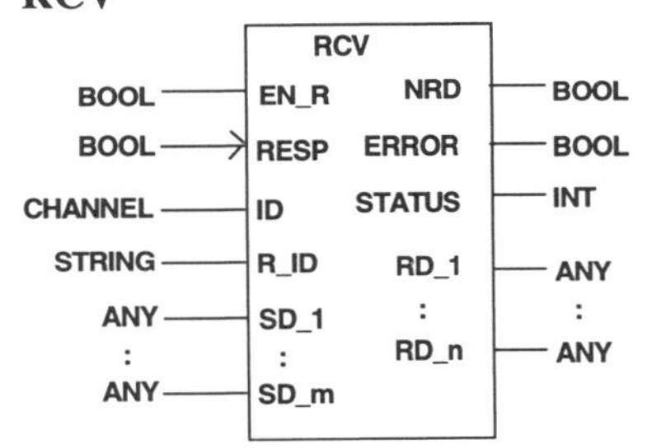

The RCV block receives a set of values RD_1 to RD_n from a remote SEND block. In response, it transmits back a further set of values SD_1 to SD_m to the SEND block. The number of SD_m inputs and number of RD_n outputs are extensible.

Note: The send data, the response data or both may be empty, i.e. where no SD_n inputs and corresponding RD_n outputs are defined.

The SEND function block instance requests that the RCV instance in a remote PLC executes an application operation and returns the result of the operation to the SEND instance. The SEND and RCV function block pair have the effect of implementing remote procedure calls.

ALARM

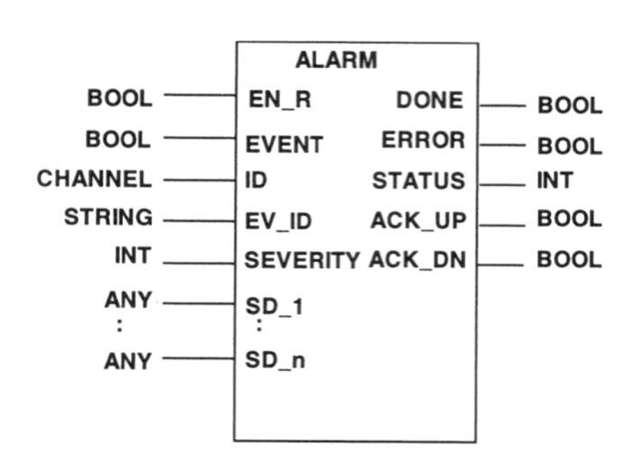

The ALARM block sends values of one or more variables to a remote PLC as identified by the channel ID. The alarm can be characterised by a severity level. The reason for the alarm or event is passed to the EV_ID input. This block expects that the remote PLC will acknowledge the reception of the alarm. The number of SD_n inputs is extensible.

The rising edge of the EVENT input is used to signal when an alarm or event condition has occurred. The falling edge of the EVENT input signals when the condition no longer exists i.e. has cleared.

NOTIFY

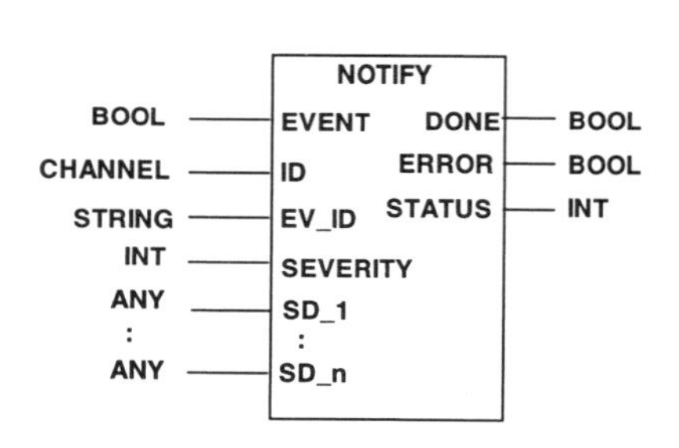

The NOTIFY block can be used to report an event or an alarm to a remote PLC but unlike the ALARM block does not require an acknowledgement.

The number of SD_n inputs is extensible.

REMOTE_VAR

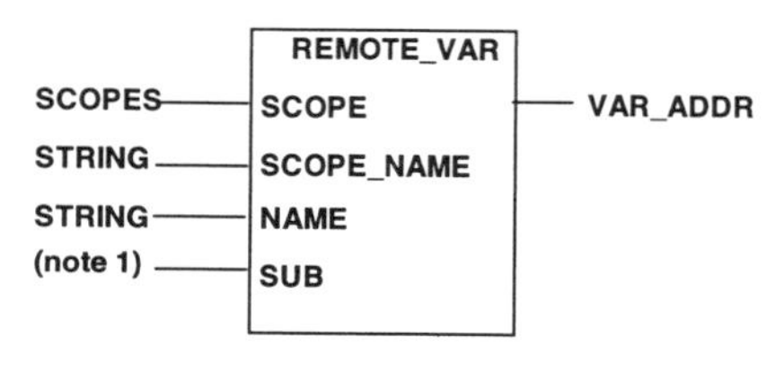

The REMOTE_VAR function block can be used to obtain an implementation specific address of a named variable. The SCOPE identifies the name scopes of the various IEC 1131-3 languages or other implementation specific scopes as defined in Table 12.4.

The input SCOPE_NAME contains the identity of the name scope for a given variable. The variable name is given by input NAME. SUB contains the name of a sub-element or an index.

> Note: Although not explicitly stated in IEC 1131-5, it is assumed that REMOTE_VAR can only reference VAR_ACCESS variables.

For example, the following input values will request the remote implementation specific address of the first element of an array SPEEDS defined as a VAR_ACCESS variable contained in program instance PRG1, held in resource RES1, defined within configuration CONFIG1.

```
SCOPE := 2,
SCOPE_NAME := 'CONFIG1.RES1.PROGRM1',
NAME := 'SPEEDS',
SUB := 1
```

The implementation specific address, e.g. the memory location, is returned in a reserved data type called VAR_ADDR. The READ and WRITE communications function blocks can reference variables using an address specified as a VAR_ADDR data type. This mechanism allows variables to be referenced efficiently using the implementation specific internal addressing scheme.

A VAR_ADDR can be regarded as a remote PLC software handle for an IEC 1131-3 variable and as such may be transitory in nature.

Table 12.4 Values of name scopes

Name scope of IEC 1131-3	Value
Configuration	0
Resource	1
Program	2
Function block instance	3
Reserved for future extensions	4
Reserved for future standards	4..9
Implementation specific	>99

12.6 MMS mapping

Part of the IEC 1131-5 standard concerns the mapping of the communications function blocks on to services specified in the MMS (Manufacturing Message Specification) standard ISO/IEC 9506-5. MMS is the application layer specified for MAP (Manufacturing Automation Protocol) - an industrial communications system initiated by General Motors (GM) in the USA in the mid 1980s.

The MMS standard defines a set of services that allow industrial devices, cell controllers and supervisory systems to exchange information over a communications network. Each service is specified as a transaction. A defined set of data items is transmitted to a remote device, which is then required to respond with one of a defined set of responses. The structure of request and response information is sufficiently flexible to allow for additional implementation specific data.

The MMS mapping defined in the standard covers all the facilities as listed in Table 12.3. Each function block will use a defined subset of the MMS services. For example, the STATUS block will use the MMS Status service. It is envisaged that the implementation of each communications block will call a defined set of MMS services. This will allow IEC compliant devices to be able to interoperate on any MMS based network.

Using MMS also makes it possible for devices on an MMS network, which are not implemented using the IEC 1131-3 standard, to communicate with IEC 1131-3 based PLCs. For example, a numerically controlled (NC) lathe on an MMS network could respond to a STATUS function block invoked within an IEC 1131-3 based PLC, provided that it supports the necessary MMS service subset.

IEC 1131-5 defines the mapping between IEC 1131-3 and MMS and data types.

Examples of some IEC 1131-5 MMS mappings:

IEC 1131-3 data type	**MMS type description class and size**
BOOL	*boolean*
SINT	*integer 8*
INT	*integer 16*
DINT	*integer 32*
LINT	*integer 64*
USINT	*unsigned 8*
REAL	*floating-point 32,8*

Note: The full MMS data type mapping is given in Table 49 of the IEC 1131-5 standard.

IEC 1131-5 also defines rules for data type compatibility, the mapping of IEC 1131-3 name scopes to the various MMS domains and the MMS transactions required to support the interaction of the IEC 1131-5 communications function blocks.

Summary

In this chapter we have outlined the main concepts defined in the IEC 1131-5 communications standard. To summarise:

- IEC 1131-5 uses concepts from the PLC languages standard - specifically access variables and function blocks.
- The status of each subsystem within the PLC, such as memory, I/O system etc., can be accessed as a standard structure at a fixed communications address.
- Standard communications function blocks are defined to support a range of facilities, such as device verification and data acquisition.
- Many of the function blocks have a server or client role. For example, a local USEND block exchanges data with a remote URCV block.
- Variables within remote PLCs can be addressed using the implementor's own addressing scheme.
- The communications function blocks are designed to map on to MMS services so that PLCs compliant with IEC 1131-5 can interoperate on an MMS based network.

Appendix 1

IEC 1131-3 keywords

This appendix contains a list of keywords used in the IEC languages. To avoid the possibility of having ambiguous language identifiers, the names of the following keywords should be regarded as reserved and not used as identifiers; see Chapter 3 for further details on identifiers.

Keywords	**Comments**
N R S L D P SD DS SL	Action block qualifiers used in Sequential Function Charts
ACTION END_ACTION	Declares an action block body
ARRAY OF	Array variable definition
AT	Used to associate variables with directly represented variables
CASE OF ELSE END_CASE	Case statement in Structured Text
CONFIGURATION END_CONFIGURATION	Declares a configuration body
CONSTANT	Attribute of a variable
EN ENO	Control of functions and function blocks in Ladder Diagram language
EXIT	To exit from an iteration loop in Structured Text
FALSE	Boolean literal (constant) for false or = 0
F_EDGE	Falling edge data type
FOR TO BY DO END_FOR	Iteration constructs in Structured Text
FUNCTION END_FUNCTION	Declares a function body
FUNCTION_BLOCK END_FUNCTION_BLOCK	Declares a function block body
IF THEN ELSIF ELSE END_IF	Conditional statements in Structured Text
INITIAL_STEP END_STEP	Declares an initial step body

Keywords	Comments
NON_RETAIN	Attribute of a variable that defines that its value is never retained when there is a powerfail[1]
PROGRAM WITH	Used to configure a program
PROGRAM END_PROGRAM	Declares a program body
R_EDGE	Rising edge data type
READ_ONLY READ_WRITE	Attributes of a variable
REPEAT UNTIL END_REPEAT	Iteration constructs in Structured Text
RESOURCE ON END_RESOURCE	Used in resource configuration
RETAIN	Attribute of a variable that defines that the value is retained during a powerfail
RETURN	Defines an early exit from a function or function block
STEP END_STEP	Declares a step body
STRUCT END_STRUCT	Used to declare a structured data type
TASK	Used to declare a task
TRANSITION FROM TO END_TRANSITION	Declares a transition body
TRUE	Boolean literal (constant) for true or = 1
TYPE END_TYPE	Declares a derived data type

[1] *Keyword added as an amendment to the 1993 edition of IEC 1131-3.*

Keywords	**Comments**
VAR END_VAR	Declares a set of variables
VAR_INPUT END_VAR	Declares a set of input variables
VAR_IN_OUT END_VAR	Declares a set of input/output variables
VAR_OUTPUT END_VAR	Declares a set of output variables
VAR_EXTERNAL END_VAR	Declares a set of external variables
VAR_ACCESS END_VAR	Declares a set of access variables
VAR_GLOBAL END_VAR	Declares a set of global variables
WHILE DO END_WHILE	Iteration construct in Structured Text
WITH	Used to associate a program or function block with a task

Note that the following should also be considered to be reserved:

- Standard data type names, e.g. BOOL, REAL
- Standard function and function block names, e.g. SIN, COS, RS, TON
- Textual operators in the Instruction List language, e.g. LD, ST, ADD, DIV
- Textual operators in the Structured Text language, e.g. NOT, MOD, XOR

Appendix 2

IEC 1131-3 amendments

This appendix summarises the main amendments to the 1993 revision of the IEC 1131-3 standard currently proposed by the IEC Task Force, IEC 65B/WG7/TF3, to go into the second edition of IEC 1131-3 planned to be published in 1999.

At the time of writing this appendix, autumn 1998, these amendments have been incorporated in the second edition of IEC 1131-3. The draft of the second edition will be distributed to national standardisation bodies for review and comment during the spring of 1999. The second edition is planned for publication late summer 1999.

The fact that the IEC is working on new editions of IEC 1131-3 indicates that it is an emerging standard that will continue to be refined and extended as users start to apply the IEC 1131-3 concepts and languages to the development of mainstream control system software.

Typed literals

There are situations where the type of data represented by literal values, such as 12.43, 73, 2#1001, is ambiguous. This is a particular problem in the Instruction List (IL) language; for example, when a literal value is used with a load (LD) instruction, the type of data loaded into the IL register cannot be specified.

It is proposed that all literals can be prefixed with their data type in the form '<data type> #'. Examples are:

```
SINT#46         (* Short integer literal *)
LREAL#12.65      (* Long REAL literal *)
PRESSURE#43.2   (* Literal of user type PRESSURE *)
BOOL#1          (* Boolean literal *)
```

With bit string data types and integer literals, the representation format can be included in the prefix (see Section 3.6). Examples are:

```
WORD#16#FF00(* Word literal in hexadecimal *)
BYTE#2#11010011  (* Byte literal in binary *)
```

Enumerated data types

There are several deficiencies with the use of enumerated data types in the 1993 revision of the standard. As with literals, the data type of a particular enumerated value is ambiguous. For example, there may be several enumerated type definitions using the enumeration string 'ON'. It is therefore proposed that the data type of a particular enumeration literal can be specified using a prefix in the form '<data type>#'. Examples are:

```
TYPE
  VALVE_MODE: (OPEN, SHUT, FAULT);
  PUMP_MODE: (RUNNING,OFF,FAULT);
END_TYPE;
...
IF AX100 = PUMP_MODE#FAULT  THEN
         XV23 = VALVE_MODE#OPEN;
```

The syntax of Structured Text is to be extended to allow enumerated variables to be used in CASE statements, e.g.

```
TYPE
  BATCH_TYPE: ( SMALL, LARGE, CUSTOM);
END_TYPE;
VAR
   NEW_BATCH: BATCH_TYPE;
END_VAR;
...
CASE NEW_BATCH OF
  SMALL:    ...  (* Small batch *)
  LARGE:    ...  (* Large batch *)
  CUSTOM:    ... (* Custom size batch *)
ELSE
...
```

Temporary variables

With the 1993 revision of IEC 1131-3, there is no provision to create variables within programs and function blocks to hold temporary values. Values held in variables declared using the VAR construct within POUs always persist between POU invocations. Using such variables for temporary values can result in an inefficient use of memory.

There is a proposal that temporary variables can be declared using a VAR_TEMP construct. Such variables will be placed in a temporary memory area, e.g. such as on a stack, which is cleared when the POU invocation terminates. For example:

```
VAR_TEMP
  RESULT : REAL;
END_VAR;

RESULT := AF18 * XV23 * XV767 + 54.2;
OUT1 := SQRT(RESULT);
```

Type conversion functions

There are a number of BCD (binary coded decimal) data type conversion functions where the name of the function is not consistent with the data type of the initial or converted variable. An amendment is proposed to provide BCD conversion functions that include the data type of the BCD value in their name.
Examples of some new BCD type conversion functions are:

WORD_BCD_TO_UINT()	Converts a word bit string containing a BCD value to an unsigned integer.
UINT_TO_BCD_DWORD()	Converts an unsigned integer to a BCD value in a double word bit string.

Data type conversion functions will be provided for BCD values held in BYTE, WORD, DWORD and LWORD variables.

Functions of time data types

There are also conflicts with the rules for overloading functions for time and date calculations. The following new functions are therefore proposed:

ADD_T_T(...)	Adds a duration time to another duration time to produce a new duration time.
ADD_TOD_T(...)	Adds a time of day to a duration time to produce a time of day.
CONCAT_D_TOD(...)	Concatenates a date and a time of day to produce a new date and time.

Pulse action qualifiers

The pulse action qualifier P actually causes an action to execute twice, once when the action is first activated and again after the action's associated qualifier Q is cleared (see page 196). To provide a more consistent single pulse behaviour two further action qualifiers are proposed:

Qualifier	Description
P1	Single pulse action when the action qualifier becomes true (rising edge)
P0	Single pulse action when the action qualifier becomes false (falling edge)

Action control

It is proposed that the action qualifier Q can be tested within an action. This allows an action to detect when it is executing for the last time, i.e. on the falling edge of Q. The proposed behaviour is already described in Chapter 8, Section 8.7 'Action qualifiers'.

Function invocation with VAR_IN_OUT

An amendment to the second edition of IEC 1131-3 allows inputs to functions to be declared as type VAR_IN_OUT if required. This implies that variables that are passed as VAR_IN_OUT may be modified within the function body. This is similar to passing a variable by 'reference' rather than by 'value'. This feature has been added to allow functions to work more efficiently with multi-element variables, i.e. structures and arrays.

For example:

```
FUNCTION  SetTests : INT
 VAR_IN_OUT
   Tests[1..10] : INT;
 END_VAR
 VAR_INPUT
   TestValue : INT;
 END_VAR
 VAR
   I1 : INT;
```

```
    Count : INT := 0;
  END_VAR

  FOR I1 := 1 TO 10 DO
    Count := Tests[I1] + Count;
    Tests[I1] := TestValue;
  END_FOR;

  SetTests := Count;
END_FUNCTION

(* Example using SetTests() function *)
VAR
  M20_Tests[1..10] : INT;
  Alarm : BOOL := FALSE;
END_VAR;
...
IF SetTests(Tests := M20_Tests,
  TestValue := 10) > 20 THEN
  Alarm := TRUE;
END_IF;
```

The function 'SetTests' sets the value of all elements of the array 'Tests' to 10. It also calculates the original sum of all values in the 'Tests' array and returns the value as the result of the function invocation.

Support for double-byte strings

An additional data type 'WSTRING' is proposed to allow strings to be defined to hold double-byte characters. These are required for handling messages in languages with complex character sets, such as Japanese. The string handling functions will be overloaded so that they can be used with double byte strings.

RETAIN and NON_RETAIN variable attributes

The use of the RETAIN attribute with multi-element variables such as structures and function block instances is modified in the second edition of IEC 1131-3. When RETAIN is used with a multi-element variable it applies to all contained variables except those that are declared with a NON_RETAIN attribute.

NON_RETAIN is a new attribute and indicates that the variable's value is not retained during a powerfail and its default initial value is to be used after a warm restart. If NON_RETAIN is applied to a multi-element variable, then it applies to all contained variables that are not declared with a RETAIN attribute.

Use of directly represented variables in function blocks

It is proposed that variables within function block bodies can be declared with the AT attribute so that they can be fixed at particular memory or I/O locations. It will also be possible to use directly represented variables within function blocks. This has been added to allow function blocks to be used for 'modularisation' of large programs. In other words, it permits a large program to be broken down into a set of smaller function blocks. Allowing memory and I/O addresses to be embedded inside function blocks helps to simplify the overall design.

However, it should be noted that function blocks that contain directly represented variables generally will not be re-usable in other systems.

Extended initialisation facilities

One major shortcoming of the 1993 revision of IEC 1131-3 has been the lack of facilities to define configuration parameters for function blocks, programs and complete configurations. Complex function blocks, such as for an advanced PID controller, generally require a large number of configuration parameters, e.g. for defining tuning parameters, operating modes and time constants. In order to make such blocks general purpose, all configuration parameters required for the internal algorithm need to be brought out to the block's outer interface. In many cases, this results in having a function block with an overly complex interface with many inputs only required once when the block is first called.

For these reasons, the following new features to improve initialisation facilities are proposed:

Function block instance initialisation

To avoid the need to invoke a function block instance simply to define its initial input values, i.e. to set-up input values needed to configure the internal algorithm, it is proposed that initial values for inputs and internal variables can be defined directly when the function block instance is declared.

For example, setting the initial input values for an instance of the Ramp function block:

```
VAR PressureRamp :
   RAMP := ( CYCLE := T#1S,  (* Set initial values
for *)
   TR := T#30M);                 (* Cycle and Ramp
durations *)
END_VAR
```

Note: Initial values for variables given in the function block instance declaration will overwrite any default initial values defined in the function block type definition.

Configuration initialisation

Most large IEC 1131-3 configurations will generally necessitate the definition of many constants and I/O addresses in order to uniquely tailor the configuration for use with a specific PLC system or application. Users have found that there is sometimes a requirement to re-configure a proven IEC 1131-3 configuration for use on different hardware or on a different application, e.g. to change a configuration to work on a slightly different set of PLCs, or on a slightly different production line or process. In such cases, the configurations may only differ in terms of configuration parameters such as algorithm tuning parameters and the addresses of I/O points.

To support this requirement, it is therefore proposed that a 'baseline' configuration can be created. Configuration specific values can then be defined in a single construct called VAR_CONFIG. This defines specific values that are used to redefine the initial values of specified program inputs, program internal variables and function blocks.

Consider the following example:

```
CONFIGURATION  LINE_3
...
RESOURCE MACHINE1 ON PROC_486
...
   PROGRAM PRG1 : PRG1_TYPE (...
...
   PROGRAM PRG2 : PRG2_TYPE (...
...
END_RESOURCE

   (* Define Line 3 specific data *)
   VAR_CONFIG
     MACHINE1.PRG1.LINE_LENGTH : INT := 20;
     MACHINE1.PRG1.SPEED : REAL := 6.6;
     MACHINE1.PRG2.RATE1 : TON := (PT := T#1M);
   END_VAR
END_CONFIGURATION
```

This example shows how the initial values of variables LINE_LENGTH and SPEED, declared within program PRG1, and resource MACHINE1 can be initialised to specific values. RATE1, which is declared within program PRG2, is an instance of an on-delay timer (TON) that is initialised with a specific time duration for input PT. The VAR_CONFIG can contain the initial values for any variable within a program, except variables declared using VAR_TEMP, VAR CONSTANT or VAR_IN_OUT. Each variable must be identified by concatenating the resource name, program name, function block instance name and so on, down through the block hierarchy.

'Wild-card' direct addresses

The definition of direct addresses for some variables, i.e. addresses used with the AT construct, is another aspect of a configuration which may require redefinition when a configuration is modified to work on a different physical system.

For example, a sensor connected to a PLC input may be at a particular physical input channel and I/O rack position in one system but be connected to a completely different physical input channel and rack in another. In order to use the same software, i.e. the same IEC 1131-3 configuration in the two systems, it is necessary to redefine the direct address used for the input. With the 1993 revision of IEC 1131-3, changing direct addresses of variables can only be achieved by modifying various variable declarations throughout the configuration and then recompiling the configuration.

It is proposed that each variable for which the direct address is to be redefined is declared using a 'wild-card' address specified as an asterisk. Such variables are considered to be 'not located'. Examples are:

```
VAR INPUT1 AT %IX* :BOOL;(*Boolean input not located *)
VAR VALV1 AT %QW* : INT;(*Integer output not located *)
```

The location of such variables can be specified in the VAR_CONFIG construct. For example, the following statements will define the location of these variables; assume that INPUT1 and VALV1 are declared within resource RES1.

```
VAR_CONFIG
   RES1.INPUT1 AT %IX100 : BOOL;(* Locate input1 *)
   RES1.VALV1 AT %QW210 : INT;(* Locate valve 1 *)
END_VAR
```

It is assumed that initial values for the variables specified in VAR_CONFIG will be applied as the last process before creating the program object data that are downloaded into the PLC.

Note: An error will be reported when the configuration is built if any variable declared with an un-located direct address is not given a valid direct address.

Amended Instruction List (IL)

There is a proposal to improve the description of the Instruction List (IL) language. Many of the proposals are already described in Chapter 7, 'Instruction List'.

Production rules

Many of these new features will result in changes to the production rules (i.e. structure and syntax) for the textual languages. It is intended that wherever possible the changes will be 'backward compatible' and should not significantly impact on existing programming systems.

Summary

A number of other amendments are currently being considered to update the 1993 revision of IEC 1131-3. All of these changes are still under review and should not be considered part of the standard until officially published by the IEC.

Appendix 3

IEC 1131-3 frequently asked questions (FAQ)

The following list is a compilation of frequently asked questions about IEC 1131-3. The reader should be aware that it has been the author's best intention that the answers to these questions are correct. However, in due course, it is possible that changes in new editions of the standard may modify or possibly invalidate some of the answers.

- **Will IEC 1131-3 reduce the innovation of new languages and concepts for PLCs?**

The main objective of IEC 1131-3 has been to standardise existing PLC languages. There is no intention that IEC 1131-3 should reduce the development of new PLC languages. Any PLC vendor is free to provide extensions and additional languages where required. Because the standard allows proprietary function blocks to be programmed in non-IEC 1131-3 languages such as C++, it is always possible to provide extensions fairly 'seamlessly', e.g. packaged as function blocks. This is well demonstrated by IEC 1131-7 "Fuzzy control programming" which defines language extensions for implementing fuzzy logic encapsulated as function blocks.

- **Why does IEC 1131-3 have 'resources'? Is a resource just another name for a PLC?**

A resource is a general name for anything that is able to provide the appropriate access to I/O systems and services to allow IEC 1131-3 programs to execute. Normally a PLC that can execute IEC 1131-3 programs can be regarded as a single resource. However, other processors, such as a personal computer (PC), if able to support the execution of IEC 1131-3 programs may also be regarded as resources.

- **IEC 1131-3 seems to be very complicated. Is it still possible to create simple Ladder programs by users unfamiliar with IEC 1131-3?**

An IEC 1131-3 system can still be programmed as a single Ladder program if required. Programming systems may provide an option to create a simple IEC 1131-3 configuration containing one resource, one task, and one program instance of a program type. All of this could be created automatically so the user is only concerned with developing a single ladder program. Function blocks and other IEC 1131-3 constructs do not need to be used.

- **Will IEC 1131-3 languages result in applications that run more slowly and require more memory than using simple ladder?**

Attempting to implement IEC 1131-3 constructs such as function blocks on PLCs that were originally only designed to support ladder programs, will inevitably have performance and memory overheads. PLCs specifically designed with firmware to support the execution of IEC 1131-3 programs should not be noticeably slower than classical ladder based PLCs. The improvements in software structure from IEC 1131-3 should allow users to be able to write more efficient applications that will be significantly easier to maintain than monolithic ladder programs. IEC 1131-3 languages tend to run more efficiently on new 'soft PLCs' than on traditional ladder based PLCs.

- **Is it really possible to port IEC 1131-3 software from one vendor's PLC to another?**

No it is not possible simply to take an application that runs on one type of PLC and copy it over to another type. There are several problems preventing the direct porting of IEC 1131-3 software.

1. The PLC I/O systems use different addressing schemes.
2. The task scan rates supported on different PLCs vary.
3. Each PLC vendor may have implemented a different subset of IEC 1131-3 features.
4. Similarly each vendor may have different values for implementation specific parameters such as maximum array sizes, string lengths etc.
5. Finally there is not a standard file format in which to store and port IEC 1131-3 applications.

Notwithstanding these constraints, at the function and function block level, it may be possible to re-implement identical POUs on different vendors' PLCs. Textual source code for POUs developed in ST or IL can be ported between different types of PLCs.

- **Is it possible to automatically convert between IEC 1131-3 languages: for example, can a POU written using LD be viewed and edited in ST or FBD?**

This is a favourite IEC 1131-3 myth. There has never been any intention that it should be possible to convert any language into any other language. If a restricted subset of each language is used some limited portability may be possible but there are some significant problems. For example, there is no way to represent expressions involving array variables in the FBD language.

- **Can function blocks also have execution control variables EN and ENO, like functions?**

The standard is not explicit about whether function blocks may have execution control variables, e.g. for connecting function blocks within ladder rungs. However, from the IEC 1131 languages user guidelines (part 8 of the IEC 1131 standard), it is implied that for consistency, both functions and function blocks should use EN and ENO variables for execution control in ladder diagrams. It is an implementation decision whether function blocks have EN and ENO variables that can be used in the FBD language for explicit execution control. This may be useful in eliminating execution order ambiguities that might arise in FBD networks.

- **In a full graphic implementation of FBD, is there any way to distinguish between lines that cross over and lines that join?**

The graphical format of lines, crossovers and junctions in full graphic implementations of languages LD, FDB and SFC is not specified in IEC 1131-3. It is an implementation decision outside the scope of the standard how fine graphic details, such as 'line crossovers', are depicted.

- **Where are type definitions actually defined and what is their scope?**

All type definitions for data types and POUs can be regarded as outside the entire IEC 1131-3 configuration and apply to all entities within the configuration, i.e. all type definitions have global scope. This is as if all type definitions exist in a conceptual 'header file' that is pre-processed before any entity within a configuration is compiled. Extensions to IEC 1131-3 are now being considered to provide a more flexible range of type scopes. With large applications, more specific type scopes may be necessary, such as a library scope for type definitions that only apply to POUs within a specific library.

- **Do IEC 1131-3 languages enforce data type consistency?**

All IEC 1131-3 languages except IL enforce strict data type consistency, i.e. it is not possible to directly assign (or connect) variables of one data type to variables of different data types. Data type conversion functions are necessary to convert the values of variables to the appropriate type; e.g. an INT value should be converted to a REAL value before being assigned to a REAL variable. In the IL language, it is not always possible for a compiler to check that the type of value in the accumulator will match the data type of any variable to be loaded from the accumulator. With IL, run-time checks are required to ensure data type consistency.

- **When are the actions in an SFC actually executed?**

Every SFC is encapsulated in a function block or program POU. When the POU is invoked, e.g. because it has been scheduled by a task, the contained SFC is evaluated once, i.e.

- The current set of active steps is determined.
- All transitions associated with active steps are evaluated
- Actions which nominally ceased execution in the previous SFC evaluation (because their Q flag has been cleared) are executed one last time.
- All actions that are active are executed once.
- Any active steps that precede transition conditions that are true are deactivated and their succeeding steps are activated.

The encapsulating POU should be repeatedly invoked for the SFC to progress through its various steps.

- **How can the execution of all actions in an SFC be halted and the SFC be restarted?**

The execution of all actions in an SFC can be halted by suspending the invocation of the encapsulating POU: see "When are the actions in an SFC actually executed?"

If, at a later time, the POU is again repeatedly invoked, the active actions in the contained SFC will continue to be executed. The only way to re-start an SFC from its initial step and clear all active actions is for the resource containing the POU to have a 'cold start'. A jump back to the initial step is always possible using an explicit branch, e.g. back from the last step in a sequence. However, this cannot guarantee to clear any stored actions or simultaneous sequences which may have been started.

- **Can more than one variable be fixed at the same direct address using the AT construct?**

The standard does not forbid this. Allowing variables to be at the same or to use overlapping memory locations is an implementation issue. This however may invalidate data type consistency - see "Do IEC 1131-3 languages enforce data type consistency?"

- **When using the AT attribute, does the size of a memory location specified by a direct address have to match the size of the variable?**

The standard is unclear; there are two ways of interpreting the purpose of the direct address. It specifies either (*a*) the actual memory location, in which case the location size and variable type size should match, or (*b*) the starting address from which the variable will be located, in which case sizes do not need to match.

- **When are user specified initialisation values for variables used?**

User specified initial values apply to non-retentive variables both at 'cold restart' and 'warm restart', but they only apply to retentive variables at 'cold restart'. On 'warm restart' retentive variables have the same values as existed when their resource stopped executing, e.g. due to a power outage. The 1131-3 amendment allows initialisation values defined by the VAR_CONFIG construct to override type specific initial values. Therefore, VAR_CONFIG specified initial values apply at 'cold restart' or 'warm restart' for non-retentive variables, and to 'cold restart' for retentive variables.

- **Can function block instances be passed as inputs to other blocks?**

Yes, if function block FB1 is passed as an input to a second function block FB2, it is possible to invoke the function block FB1 within the body of FB2. Any input parameters of FB1 not defined in the invocation call within FB2 will take values defined by earlier invocations of FB1 made outside FB2. Function block instances should be passed as VAR_IN_OUT parameters as otherwise the values produced within the internal invocation will not be preserved.

- **Are the assignments of inputs and outputs to programs at the resource level fixed or can they change dynamically?**

They are fixed. Programs are the highest level IEC 1131-3 programmable organisation units. It is not possible to have language statements outside a program. In a resource declaration it is only possible to assign input and output variables to the program interface. If it is necessary to change the inputs used by part of a program, then all inputs concerned should be passed to the program and any dynamic selection of particular inputs should be done within the program body, e.g. switching between two different sets of sensors.

Appendix 4

IEC 1131-3 future developments

Considerable progress in the adoption of IEC 1131-3 languages has been made in the three years since this book was first published in 1995. At that time, the PLCopen trade association promoting IEC 1131-3 products had a membership of about 40 companies and organisations. Now in the spring of 1998, PLCopen's membership has grown to almost 100 indicating that interest in IEC 1131-3 is rapidly increasing. Most of the major PLC manufacturers now have an active interest in IEC 1131-3. Originally the main interest in IEC 1131-3 was centred in Europe, but interest is now truly international with involvement of companies from the US, Japan and Europe.

Since 1993 a wide range of automation products have appeared on the market that have adopted some aspects of the IEC 1131-3 standard; these range from programming support software for traditional ladder based PLCs, through to 'soft PLC' products, and programmable SCADA.

Worldwide, the market for PLCs and automation products is still showing healthy growth. Related to this growth in automation, there is also going to be a considerable increase in the demand for software. It is estimated that the European market alone for automation and process control software will be at least $2500 million by 2001. It is therefore likely that the demand for IEC 1131-3 based products is also going to increase dramatically.

As companies started to develop products complying with the 1993 edition of IEC 1131-3, small inconsistencies and errors in the standard were uncovered. The IEC task force, 65B/WG7/TF3 developing IEC 1131-3, categorised these as being either corrections to the published standard — generally typographical errors, or amendments, which require minor technical modifications. All corrections and amendments to the 1993 edition have now been collated and reviewed by international standardisation bodies. There is a good chance that these will be incorporated into the second edition of IEC 1131-3 which should be published towards the end of 1998 or early 1999.

Looking forward, the third edition of IEC 1131-3 is likely to have significantly more technical changes. The exact nature and format of these changes are still being considered but the main issues to be addressed are likely to be:

- Harmonisation with the function block standard IEC 1499;
- Additional features for safety critical software;
- Support for working with devices operating across Fieldbus networks;
- Extensions to bring IEC 1131-3 closer to Object Oriented languages.

The most significant new direction for PLC applications that is foreseen in the near future is the development of truly distributed applications. No longer will system engineers be concerned about an application that runs solely on a single PLC. Future systems are most likely to involve distributed applications that run over a variety of network topologies. For example, large automation applications are likely to communicate across the Internet, factory local area networks (LANs) and proprietary networks.

As applications become more distributed, it will become more difficult to identify the true boundaries of the application. This is synonymous with the structure of the Internet World Wide Web where it is sometimes difficult to identify the boundaries of a Web site - as one site may have links with other sites and become dependent on them. Handling complex distributed automation applications and dealing with issues such as version control and on-line maintenance will prove to be particularly challenging.

To achieve large scale distributed applications for industrial control, and yet enable the creation of flexible systems that can be re-engineered as industrial and business needs change, will require a completely new approach to software design - a new technology based on the interaction of distributed objects. There are several software technologies already well advanced that are set to have an influence in this area. CORBA (Common Object Request Broker Architecture) is a new standard for designing distributed objects that is being developed by a consortium of leading software vendors - the Object Management Group (OMG).

OPC (OLE for Process Control) based on Microsoft's OLE/COM (Object Linking and Embedding, Common Object Model) technology will allow software in the form of software components to interoperate regardless of where they are located, be it in a remote industrial controller in a blast furnace or in the PC of the production manager's office. Internet technology using Java and the World Wide Web is also being considered for the development of software components for manufacturing systems. There are even industrial devices emerging, such as smart valves that are able to execute embedded Java code directly.

The industrial community has long been aware that the ready interconnection of software components, such as in the form of function blocks, will have major advantages especially for end-users. These advantages will include improved software productivity through re-use of standard solutions and improved design flexibility by being able to plug-and-play software and devices from different vendors. So far, the new standards all enable 'technical integration' of distributed components, but the next major hurdle is 'semantic integration'. We may be able to link and exchange data between software in a remote industrial controller and a control algorithm running in a PC, but will the connection be meaningful?

What is the IEC 1499 Function Block standard?

The IEC is currently developing a new standard IEC 1499 that defines how function blocks can be used in distributed industrial process, measurement and control systems.

As we have seen from IEC 1131-3 applications, function blocks are a well-established concept for defining robust, re-usable software components. A function block can provide a software solution to a small problem, such as the control of a valve, or control a major unit of plant, such as a complete production line. Function blocks allow industrial algorithms to be encapsulated in a form that can be readily understood and applied by people who are not software specialists. Each block has a defined set of input parameters, which are read by the internal algorithm when it executes. The results from the algorithm are written to the block's outputs. Complete applications can be built from networks of function blocks formed by interconnecting block inputs and outputs.

It is anticipated that the Application Layer part of the Fieldbus communications stack will provide the software interface to allow remote function blocks to interoperate over Fieldbus. However, IEC 1499 is being developed as a generic standard that is also applicable in other industrial sectors, in fact wherever there is a requirement for software components that behave as function blocks, such as in building management systems.

IEC 1499 defines a general model and methodology for describing function blocks in a format that is independent of implementation. The methodology can be used by a system designer to describe how a distributed control system is constructed by interconnecting function blocks on different processing resources.

Summary

During the next few years the main focus of IEC 1131-3 standardisation work is clearly going to concern language concepts and extensions for building distributed applications. Recent developments in the use of distributable software components and concepts such as re-usable software as function blocks are likely to have a major impact in this area.

The IEC and the international standardisation bodies have done a great deal of work to create standards in this area. In the long term, the unification of software languages and architectures for distributed industrial control applications will bring major economic advantages to end-users and system builders alike.

Bibliography

Cox B.J. (1986). Object Oriented Programming, An Evolutionary Approach. Addison Wesley.

International Electrotechnical Commission (1992). Programmable controllers Part 1 General information, IEC 1131-1, IEC Geneva.

International Electrotechnical Commission (1993). Programmable controllers Part 3 Programming Languages, IEC 1131-3, IEC Geneva (also British Standard BS EN 1131-3 : 1993).

International Electrotechnical Commission (1988). Preparation of function charts for control systems, IEC 848, IEC Geneva.

International Electrotechnical Commission SC65B/WG7/TF7 (December 1993). Programmable controllers Part 5 : Communications, IEC Task Force working draft.

Norme Francaise Enregistree (June 1982). Function Chart 'Grafcet' for the description of logic control systems. NF C03-190.

Parr B. A. (1993). Programmable Controllers, An Engineer's Guide. BH Newnes.

Related papers

Boudreaux J.C. (1988). Requirements for global programming languages, MAPL Symposium on manufacturing application languages, Ottawa.

Juer J. and Hughes I.P. (1991). IEC 65B Control Languages - a practical view. IEE Conference, Software Engineering for Real Time Systems, Cirencester, England.

Lewis R.W. (1993). The new IEC 1131 standard for programmable controller languages brings the benefits of 'Open Systems'. IEE International Workshop, Software Engineering for Real Time Applications, Cirencester, England.

PLCOpen

For further information on companies and products using IEC 1131-3 contact:

PLCopen
PO Box 2077
5300 CN Zaltbommel
Netherlands

Index